# Lecture Notes in Earth Sciences      54

Editors:
S. Bhattacharji, Brooklyn
G. M. Friedman, Brooklyn and Troy
H. J. Neugebauer, Bonn
A. Seilacher, Tuebingen and Yale

W0042860

Hans Zijlstra

# The Sedimentology
# of Chalk

 Springer

Author

Dr. Johannes J. P. Ziljstra
Institute of Earth Sciences
Department of Sedimentology
Utrecht University
Budapestlaan 4, P. O. Box 80.021
3508 TA Utrecht, The Netherlands

"For all Lecture Notes in Earth Sciences published till now please see final pages of
the book"

ISBN 978-3-540-58948-8        ISBN 978-3-540-49153-8 (eBook)
DOI 10.1007/978-3-540-49153-8

CIP data applied for

Originally published by Springer-Verlag Berlin Heidelberg New York in 1995

Typesetting: Camera ready by author
SPIN: 10482911        32/3140-543210 - Printed on acid-free paper

# PREFACE

The sedimentology of Chalk describes processes that caused the rhythmic vertical variation in grain size, structures and authigenic mineral concentrations in Late Cretaceous and Early Tertiary, subtropical, shallow marine, fine-grained, detrital bioclastic carbonates of northwest Europe. In particular, attention is paid to the sedimentology of the Tuffaceous Chalk of Maastricht (The Netherlands), a coarse-grained variety of Chalk that resembles the Chalk (coccolithic mudstones) as well as modern shallow marine carbonate sands.

Numerical models are presented that enable the simulation of the genesis of flint nodule layers, hardgrounds and complex wavy bedded sequences, such as the K/T boundary sequence of Stevns Klint (Denmark). The aim of this book is to show how depositional and early diagenetic features, which are observed in small-scale Chalk outcrops, can be used to reconstruct the large-scale dynamics of the northwest European continent during the Late Cretaceous and Early Tertiary.

This book could not have been written without the support of others. In particular I like to mention Ir. W.M. Felder (R.G.D. Heerlen), who introduced me to the Chalk of Maastricht many years ago and who provided the foundation for this work; NWO (AWON) for financial support of project no. 751356015; the Shell Company, Dr. W.J.E. van de Graaff (KSEPL); the management of quarries North, Romont, Curfs, Blom, Nekami and B.J. Rijk, engineer of quarry ENCI; Frijns Maintenance (Cadier en Keer, The Netherlands); The Comparative Sedimentology Division, Dr. P.L. de Boer, M. Reith, T. Zalm; the Institute of Earth Sciences of the University of Utrecht, J.W. de Groot (rock preparation), F.J. Quint, B.J.M. Benders (audiovisual services), ir. T.G. Djie-Kwee (general service laboratory), A.W.H. Jansen-Corbeek (library); Dr. M. Terlou (Department of Image Analysis & Processing Institute of Biology, Utrecht University); Dr. M.J.M. Bless, Ir. P.J. Felder (Natuurhist. Museum of Maastricht); Prof. Dr. J.E. Meulenkamp, and in particular Prof. Dr. R.G. Bromley, Prof. Dr. H. van Gemerden and Prof. Dr. C.H. van der Weijden for critical reviews of Chapter 3; Prof. Dr. W. Schlager, Prof. Dr. R.D. Schuiling and Prof. Dr. A. Seilacher for thoroughly reviewing the entire thesis and for their constructive remarks; Prof. Dr. F. Surlyk for the instructive excursion to Stevns Klint (Denmark); Prof. Dr. J.M. Hancock and Dr. A.S. Gale for pleasant discussions on Chalk; C.R. Geel for the introduction into the art of computer programming and for computer facilities; Prof. Dr Th.W. Ruijgrok and M. Ruijgrok for provision of literature, computer facilities and for pleasant discussions on cellular automata; Dr. H. Brinkhuis for passionate discussions about paleoecology; Dr. J.H. Baas, Dr. N. Molenaar, A.P. Oost, D. Pi Pujol, Dr. G. Postma, Dr. M.A. Pool and Dr. G.J. Weltje for being supportive and comprehensive colleagues. Above all I would like to thank my parents Hans and Hannie, aunt Mia, my brother Peter, my sisters Sandra and Barbara, my wife Hedia Bessaïs and my friends, Kees Geel, Frederika Prak, Sjir Renkens and Thijs Ruijgrok who where always there when needed. Last but not least, I would like to thank Poppe de Boer for his trust, his patience and his important contribution to the research and the writing of this book.

Finally, I very much appreciated the cooperation with Dr. W. Engel and Springer-Verlag.


Utrecht, September 1994                                                        Hans Zijlstra

# CONTENTS

# Summary

The Tuffaceous Chalk of South Limburg (The Netherlands), a friable, porous, bioclastic carbonate sandstone, became subject of scientific interest in 1770, when workers found a large skull of *Mosasaurus camperi* in the subterranean quarry of Mount St. Pieter near Maastricht. After Dumont (1849) had considered the Tuffaceous Chalk of Maastricht to represent the youngest Cretaceous deposits, many investigations were carried out concerning the taxonomy of the fossils and the bio- and lithostratigraphy of the type section of the subtropical shallow marine deposits of the "système de Maestricht", exposed at Mount St. Pieter.

The Tuffaceous Chalk is part of a 200 m thick Late Cretaceous-Early Tertiary, transgressive-regressive succession of dm-m thick, laterally continuous layers. The basis consists of Santonian-Campanian (glauconitic) quartz sands and smectitic clays, which cover the abraded and karstified Paleozoic sediments of the gently dipping, block-faulted northern flank of the Ardennes Massif. The middle part is formed by Campanian-Maastrichtian coccolithic mudstones (Chalk) with flint nodule layers that gradually change upwards into the upper part that consists of Maastrichtian-Danian bioclastic sandstones (Tuffaceous Chalk) with hardgrounds.

In this thesis attention is paid to the sedimentology of the Maastrichtian-Danian coarsening upwards (Tuffaceous) Chalk sequences of Maastricht, the Gironde Estuary (France) and Stevns Klint (Denmark). The vertical rhythmic variation of grain size, structures and authigenic mineral concentrations has been measured and analyzed and is explained using numerical models that allow the simulation of the genesis of bedding in (Tuffaceous) Chalk.

It is concluded that the (Tuffaceous) Chalk was deposited at rates of cm-dm per 1000 years and that the regular bedding reflects the influence of cyclic variation of storm intensity, as a result of periodic variations in the Earth's orbital elements. The cyclic changes of climate and oceanographic conditions influenced the deposition rates, depth of storm reworking and rates of mineral authigenesis in redox zones of bacterial metabolism below and parallel to the sea bottom. As an example, the genesis of the Ir-rich clay at the K/T boundary of Stevns Klint is explained with the model developed for the reconstruction of the dynamics of the depositional-early diagenetic environment from the bio- and lithostratigraphy of (Tuffaceous) Chalk sequences.

**Chapter 1** - The geology of South Limburg and the Late Cretaceous Chalk and related sediments were first described *in extenso* by Binkhorst van den Binkhorst (1859). By resuming his monograph an introduction to the sedimentology of the (Tuffaceous) Chalk of South Limburg is provided.

**Chapter 2** - Many researchers have investigated the Chalk since the 19th century. The most relevant results from the literature are discussed.

**Chapter 3** - The Chalk contains a considerable amount of non-skeletal minerals, such as glauconite, pyrite, carbonate cement and silica. These minerals occur concentrated in concentric zones around exceptionally deep burrows. They formed in aerobic to anoxic redox zones as a result of bacterial metabolism. The authigenesis of these minerals is discussed.

**Chapter 4** - Silica concretions (flint) are common in Chalk and they formed after deposition. A numerical model for flint nodule genesis is presented. The model describes the relation between the morphology of the nodules and the production and distribution of early diagenetic authigenic silica.

**Chapter 5** - The variation of silica concentration in a Chalk sequence characterised by planar, parallel flint nodule layers has been measured in detail using automatic image analysis. A numerical depositional model is presented explaining how authigenic silica was concentrated in a redox zone below the sediment surface during slow deposition. The rhythmic vertical variation of the silica concentration is related to oceanographic changes due to orbital variations. In reverse, this relation is used to reconstruct the depositional environment.

**Chapter 6** - Layers of different authigenic minerals form complex cycles at the boundary between fine-grained Chalk and coarser grained Tuffaceous Chalk. Variations in the genesis of glauconite and carbonate cement layers (hardgrounds) are related to orbital forcing.

**Chapter 7** - Coarse-grained Tuffaceous Chalk is characterised by undulous, disconformable and laterally discontinuous layers. A numerical model is presented that explains the relation between wavy bedding and early diagenetic authigenesis, in particular the lithification and genesis of hardgrounds in (Tuffaceous) Chalk.

**Chapter 8** - The principles of Chalk sedimentology, developed so far, are tested and used to explain the sedimentology, and in particular the anomalous high iridium concentration at the boundary between the Cretaceous and Tertiary in the classical Chalk sequence at Stevns Klint (Denmark).

**Chapter 9** - The genesis of wavy bedded cycles in chalk which lithifies simultaneously is simulated with a numerical model.

# Introduction

Sedimentologists study deposits of particles that once moved along the Earth's surface and formed sediments. The particles and the sediments are very diverse. Particles can be ions or molecules that precipitate from solution, while precipitates form salt layers at the bottom of a hyper-saline lagoon, they may be quartz sand grains that have been eroded from a weathered granite, transported by and deposited in a river or they may be rocks up to the size of a house, avalanched downslope and deposited at the foot of a steep mountain range. A fossil sedimentary succession of salt, sand and rock breccia may thus indicate the change of an environment of deposition from, for instance, a dry hot arid coast via a wet temperate fluvial plain to a glacial alluvial plain, caused by uplift of the Earth crust in relation to continental drift.

Comparative sedimentology makes use of the actualistic principle. It compares the presently forming sediments with fossil deposits and it presumes that the paleo-environment of deposition was comparable to the present one, as long as the present and the fossil sediments are similar. The reconstruction of successions of depositional environments in space and time contributes to a better understanding of the Earth's dynamics, and it facilities prediction of the spatial-temporal position of valuable sedimentary deposits, containing fossil fuels or precious metals.

The actualistic principle as defined above does not hold for Chalk because there are no recent subtropical shallow marine environments that are equivalent to the extensive shallow marine subtropical environment of Chalk deposition, that covered a large part of the Eurasian continent during the Late Cretaceous (100 Ma - 65 Ma BP). The principles of actualism are restricted to the universal laws that are valid at any time and place and the physical, chemical and biological laws that hold in any shallow marine subtropical sea. The variation of the type and the distribution of biogenic detrital grains and mineral precipitates in the various Chalk sequences ideally has to be explained with a model of the depositional environment that consists of a set of fixed rules and a number of variable parameters. The only requirement is a consistency with the (universal) laws for shallow marine subtropical environments.

Before focusing on the sedimentology of the Chalk, I resume the principles of the sedimentology of (sub)tropical shallow marine sedimentary environments and I elaborate on the logic required to understand the complex processes involved.

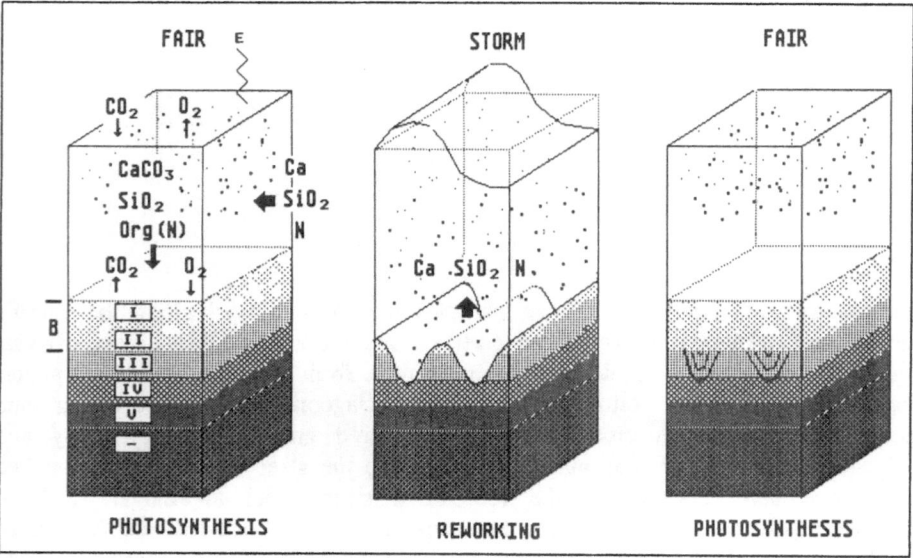

Figure 1 - Productional, depositional and early diagenetic conditions in an extensive (sub)tropical shallow sea. Fluxes of light energy (E), dissolved terrigenous matter ($Ca^{2+}$, $SiO_2$, (N)utrients), organic tissue (Org(Nutrients)), skeletal minerals ($CaCO_3$, $SiO_2$) and gasses ($O_2$, $CO_2$) during fair weather (left and right) and storms (middle). I-IV: Sediment mixing in the zone of bioturbation (B) and bacterial decomposition of organic matter, dissolution of skeletal minerals and authigenesis in different redox zones below and parallel to the sediment surface.

## The bio-chemical sedimentology of Chalk

In the centre of an extensive shallow (sub)tropical sea the water is clear and terrigenous matter is almost exclusively present as dissolved ions and molecules (Fig. 1). Unicellular algae are common and they use light energy during photosynthesis, forming organic tissue from dissolved carbondioxide, water and nutrients (P, Fe, S ...), while producing oxygen and precipitating a skeleton of carbonate or silica.

Unicellular and multicellular organisms predate on the plants and on plant eaters, using oxygen in order to burn the organic matter and to liberate the energy that is stored in organic tissue. Dead organisms sink to the sea bottom and form a sediment of skeletal carbonate/silica and organic tissue.

Multicellular organisms inhabit the sea bottom and remove organic matter from the pore space or digest the sediment and the organic matter, while mixing the sediment and leaving a trace (ichnofossil) as they move and secrete the digested sediment. Also the multicellular infauna has to respire oxygen in order to burn the digested organic matter and consequently its activity decreases towards deeper

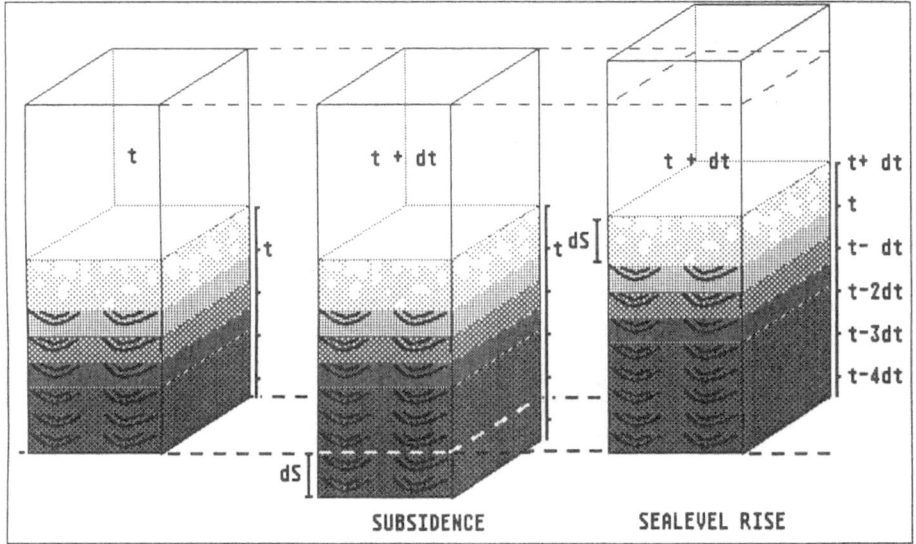

Figure 2 - Genesis of a Chalk sequence without vertical variation of lithology. Productional, depositional and early diagenetic conditions do not change. Water depth remains constant as the deposition rate (dS) equals the subsidence rate or the rate of sea-level rise. Note that the sequence does not allow a distinction between subsidence and sea-level rise.

levels in the sediment where most of the dissolved oxygen has already been used.

Below the oxygenated zone of bioturbation, organic matter is oxidized by bacteria that remove the required oxygen from oxides. With increasing depth below the sediment surface, different oxides are exhausted successively, until the least reducible oxides remain. Several redox zones can thus be distinguished, that are characterised by particular chemical conditions and by the dissolution of specific detrital biominerals and the precipitation of specific authigenic minerals.

## The physical sedimentology of Chalk

Commonly a tidal wave passes along the environment of deposition twice a day, and concurrently sea level fluctuates. Contrary to narrow, deep straits and estuaries, tidal currents in extensive shallow seas are weak and sea-level fluctuations are small. However, during storms, when the water surface is sheared by strong winds, waves are generated and oscillatory currents along the sediment surface cause resuspension, transport and sorting of considerable amounts of sediment (Fig. 1).

During the waning of storms and the subsequent decrease of wave motion the restored, the shallower part of the storm layer is destroyed by bioturbative mixing.

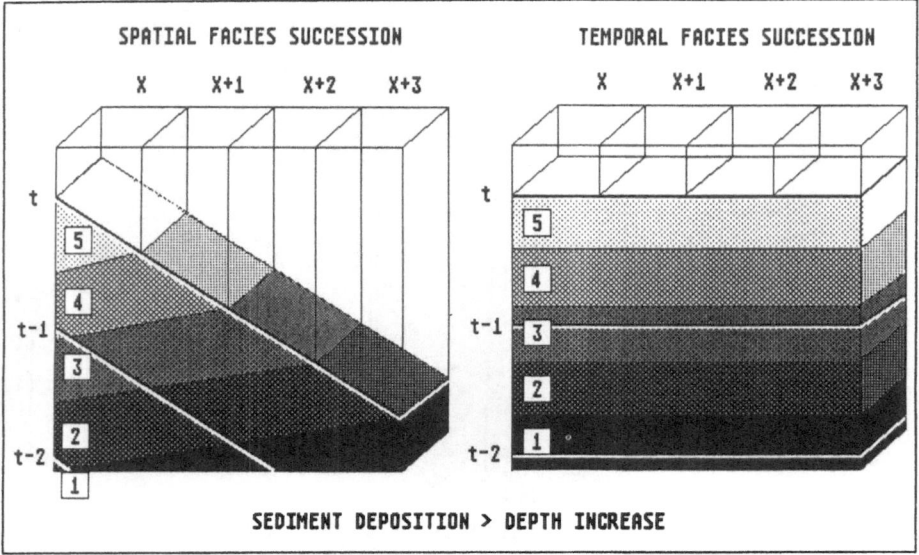

Figure 3 - Walther's law. Subsidence rate or rate of sea-level rise (depth increase) is smaller than the deposition rate. Shallow to deep facies are present simultaneously and move laterally while the marine basin is filled (left) or deep to shallow facies occur through the entire basin and succeed each other in time, while the basin is filled (right). The succession, reflected in a lithologic succession (1-5), can only be attributed to one or a combination of both cases, if the time surfaces (t) can be recognised.

The deeper part may be preserved and becomes again subject to early diagenetic reactions in the different redox zones.

## Sequences of Chalk

If the sediment production rate, the sediment composition, the conditions of (re)deposition and the conditions of early diagenesis do not change, a sequence of invariable lithology is produced, provided that the deposition rate equals the rate of subsidence and/or sea-level rise and water depth remains constant (Fig. 2).

Any deviation from such equilibrium conditions will result in the deposition of a sequence with a variable lithology. In reverse, vertical variation of the lithology informs us about changes of the sedimentary environment, although always incompletely. For instance, a lithologic variation that indicates a decrease of water depth does not provide information about the cause of the shallowing. Shallowing may be due to either a sea-level fall, a rise of the subsurface or an increase of the rate of production and deposition of sediment.

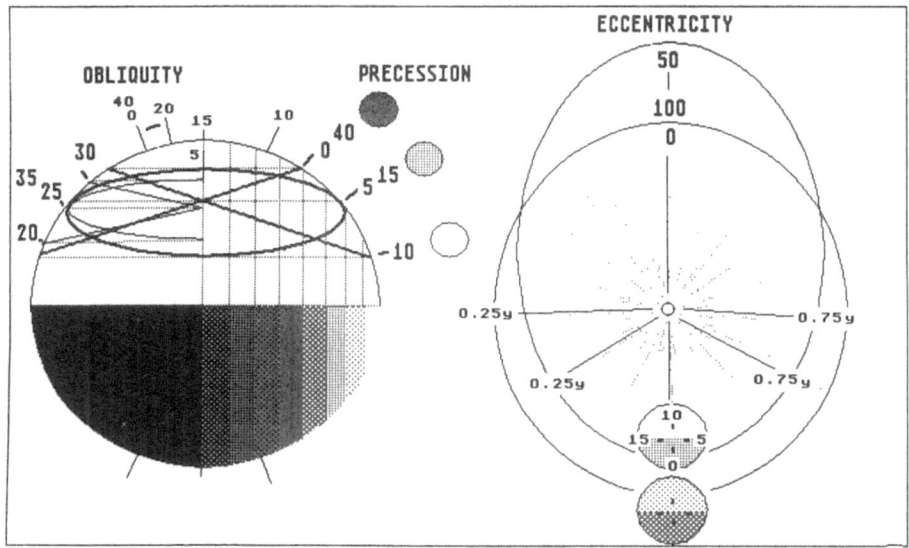

Figure 4 - The periods of the variation of the Earth orbital parameters are of the order of 100 ka for the ellipticity of the orbit around the sun (eccentricity), 40 ka for the tilt of the Earth's axis (obliquity) and 20 ka for the spinning of the Earth's axis (precession).

The intensity and distribution of solar radiation (insolation) over the year at different latitudes varies in phase with the variation of the Earth's orbital parameters.

At a northern latitude, at the year 0, when the Earth is closest to the Sun, the midday insolation is at a minimum when the Earth's axis south pole points towards the Sun and the obliquity is at a maximum. At 10 ka, when the Earth is closest to the Sun, the Earth's axis has spun 180°. Then the north pole points towards the sun and the midday insolation is at a maximum. At 20 ka, the precession has been completed, the tilt of the Earth's axis is now at a minimum and the midday insolation higher than at 0 ka. At 50 ka, the midday insolation reaches an absolute maximum due to the maximum eccentricity. The Earth is closest to the Sun during the northern hemisphere summer and a short hot solar summer is followed by a long cold solar winter.

## Walther's law in Chalk

The dynamics of the sedimentary environment in an extensive shallow sea can, to a certain extent, be reconstructed from the lithologic variation within a sequence. A number of sequences distributed over a larger area can inform about the variation of the spatial distribution of sedimentary processes in time if the instantaneous sediment surfaces (time surfaces) can be recognised. For example (Fig. 3), a shallow marine environment with a sediment production rate that exceeds the rate of subsidence and/or sea-level rise is either characterised by the lateral migration of simultaneously present, shallow to deep facies or by a succession of deep to shallow facies in time. In both cases the same lithologic succession is formed and a

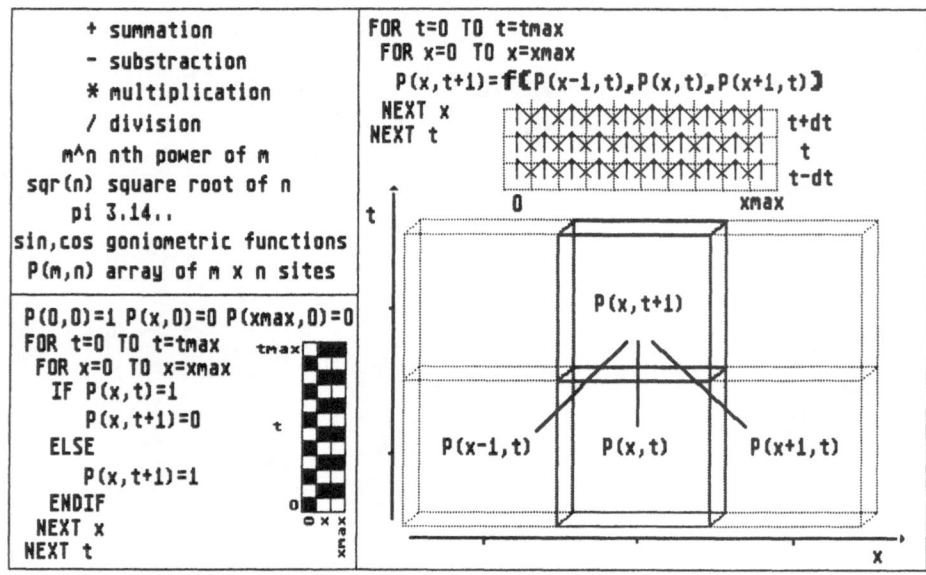

Figure 5 - Computing complex processes involving the simultaneous motion of many different masses. Several symbols used in the following chapters are explained (upper left). The description of a simple cellular automaton (lower left) is read as: Initially the property P(x,t) of the cells x=0 to x=xmax at t=0 is either 0 or 1. During steps dt, the cells x=0 to x=xmax are changed according to the rule that changes P. If P=1 in cell x at t, then P=0 in cell x at t+1 and else, if P=0 in cell x at t, then P=1 in cell x at t+1. The procedure is repeated for successive t until t=tmax.
Cellular automata used to simulate natural processes have to take into account the conditions in cells that neighbour the cell that is updated. Again for all cells x=0 to x=xmax, the property P of cell x at t+dt is calculated from the properties of cells x-1, x, x+1 at t, for t=0 until t=tmax. Note that different rules have to be defined for the boundary cells x=0 and x=xmax, unless the right and left boundary cells are connected [cell(0,t)=cell(xmax,t), continuous space].

distinction can be made only if time surfaces, that represent instantaneous sediment surfaces, are defined.


## Orbital forcing in Chalk

Chalk has been deposited at rates of several cm to dm per 1000 years. As a result of bioturbative mixing, daily (tidal) and annual (seasonal) sediment surfaces have hardly been preserved. Periodic changes of climate result from (quasi)periodic variations of the Earth orbit (Fig. 4).

These periods are of the order of 20,000 (precession), 40,000 (obliquity) and 100,000 (eccentricity) years. Considering the effects which the orbital parameters have on climate, oceanography and sedimentation, a rhythmicity is expected, that can provide an excellent chronostratigraphy.

## Computing the sedimentary processes of Chalk

The distribution of particulate matter in a Chalk sequence is the result of complex processes that involved erosion/dissolution, advection/diffusion and deposition/precipitation in space and time. Processes that are characterised by the simultaneous motion and interaction of many masses can be understood qualitatively, but in order to test such understanding on consistency and truth, a suitable language is required.

The computable language of cellular automata appears to be appropriate and has been shown to successfully describe complex processes such as the motion of gas atoms and fluid molecules. A cellular automaton is a logic rule that operates on a finite cell space. As an example, one may consider a rather unrealistic but simple cellular automaton that defines the redistribution of "mass" in a one-dimensional cell space in time (Fig. 5). In order to describe the complex dynamics of the motion of masses that led to the genesis of the Chalk, more complex cellular automata will be used, expressed in Boolean logic.

## Aim of this thesis

This thesis aims at the description of the processes that defined the sedimentology of the Chalk. The Chalk outcrops are relatively small with respect to the extensive occurrence of Chalk in the sub-surface. The gradual lateral change of the lithofacies over kilometres distance requires a detailed understanding of the small scale vertical variation of the lithology in the outcrops of the Chalk. Attention is therefore focused on the vertical variation of the depositional and early diagenetic features that define the bedding of the Chalk. The proper reconstruction of the variation of the paleo-environmental conditions in time at different locations will contribute to a better understanding of the large-scale basin-wide lithofacies variation and the dynamics of NW Europe during the Late Cretaceous.

Introductions to the various subjects can be found in: Carbonate sedimentology (Bathurst, 1971); Early diagenesis (Berner, 1980); Physical sedimentology (Allen, 1984); Orbital forcing (Berger, 1988) and; Cellular automata (Wolfram, 1986).

# 1 The Geology of South Limburg in 1859

## Introduction

The oldest known interest of Man in the rock that forms the subsurface of South Limburg, The Netherlands, dates 6000 years back. It concerns glassy black nodules of cryptocrystalline quartz, so called flint. Flint was used during the Stone Age for the production of tools and weapons, and it was excavated in a small mine at the foot of a hill near Rijholt. There, the nodules of highest quality occurred in layers within a friable, fine-grained and porous carbonate rock (Chalk).

Prehistory ended in 54 B.C., as Roman soldiers and civilians under the reign of emperor Gaius Julius Caesar entered the area between the rivers Meuse and Rhine. The Romans named the city of Maastricht "Mosae Traiectus", a bridge over the river Maas, and they introduced the art of building with stone. They started the excavation of rocks that occur above the already mentioned Chalk with flint. This younger rock is somewhat coarser grained and it contains no flint, but is equally soft, and could be manufactured with a knife.

In the subsequent centuries, several million m³ of this rock was sawn from a 7 metre thick layer within the mount of "St. Pieter" south of Maastricht (Fig. 1.1). Within this hill a labyrinthine cave complex of 300 kilometres length was created. It frequently served as a shelter for the persecuted, as witnessed by inscriptions dating from former ages, and as a grave for those who lost their way and died of desiccation (Faujas St-Fond, 1799).

Near one of the cave entrances (Fig. 1.2) a large scull of a monstrous "crocodile" was found in 1770. This attracted the attention of "learned men" (Camper, 1786), and provoked the scientific research of the geology of South Limburg. In the beginning merely the skeletal remains of the ancient life forms were collected. Most fossils were shells of snails, oysters, clams and corals. The fossils indicated the former presence of a subtropical sea at Maastricht.

A first monograph on the geology of the area between the cities of Maastricht and Aachen was published by Binkhorst van den Binkhorst (1859). After a study of eight years, during which he visited the field and discussed with "famous geologists" of that time, he presented an accurate description of the different rocks

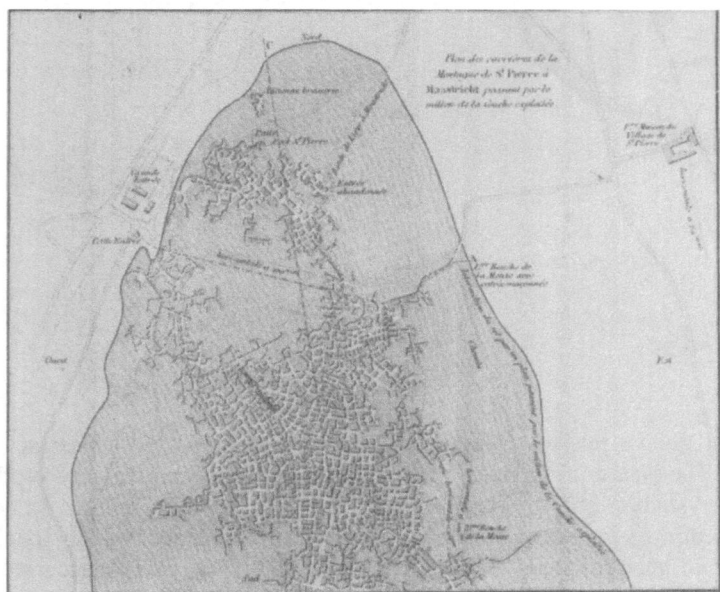

Figure 1.1 - Labyrinthine cave complex formed during the exploitation of soft, porous Tuffaceous Chalk for building purposes. The complex occurs in the one km-wide mount St Pieter just South of Maastricht (Binkhorst van den Binkhorst, 1859).

and the fossils within them. Many problems which he addressed are still subject of research today, and most of his solutions have been generally accepted and form the foundation for modern studies. It seems therefore inevitable to introduce the subject of this thesis by first resuming the most relevant parts of the "Esquisse Géologique et Paléonthologique des couches Crétacées du Limbourg" (Binkhorst van den Binkhorst, 1859).

## 1.1 Stratigraphy

According to Steno (Stensen, 1638-1686), regularly alternating stripes of different rock, exposed at the surface, are no decoration, but are successive layers, of which the underlying layer is always older than the overlying one. Such layers have been formed horizontally and if found in a steep and inclined position, a force has been acting on the layers after their formation. The rocks that are thus found at the surface, have not been formed at once, but in a succession that reflects time. Rocks were described and their occurrence and distribution were mapped.

13

Figure 1.2 - The discovery of the skull of *Mosasaurus camperi* at the entrance of the "St Pietersberg" cave complex (1770). The skull was bought by a local collector, claimed by the church who owned the overlying land, hidden during the French occupation, betrayed for 600 bottles of wine and transported to Paris by Napoleon Bonaparte. Today it is exposed in the "Musée d'Histoire Naturelle" of Paris (Faujas St-Fond, 1779).

The superposition was used to define the relative ages of the various rock layers that were distinguished. By depicting geographical distribution, properties and age of the rock, a geological map was thus constructed.

Binkhorst walked extensively through the area between Maastricht in the west and Aachen in the southeast, and he visited the many small exposures in order to construct the geological map of South Limburg (Fig. 1.3).

The soil was very fertile and the strong vegetation did not allow the study of the deeper subsurface. At most places pebble layers of Pleistocene rivers, a blanket of Holocene windblown dust or loess, or remains of strongly weathered rock covered the older layers of interest.

Few natural exposures were present, mostly along the borders of small rivers. For artificial exposures one had to dig several metres. Some boreholes provided information about the deep subsurface. These were, however, not drilled for the study of the Cretaceous sediments, but for the extraction of water and some for the exploration of coal, which in those days had just become of great economic importance.

Binkhorst distinguished 7 Cretaceous rock units, characterised by a particular petrology and, more often, by a typical fossil content. The different rocks, that always consist of particles, once transported by air or by water and subsequently deposited to form the sediment, had already been investigated, described and named by previous students of the lithostratigraphy (Le luc, 1799; Dumont, 1849) (Fig. 1.4).

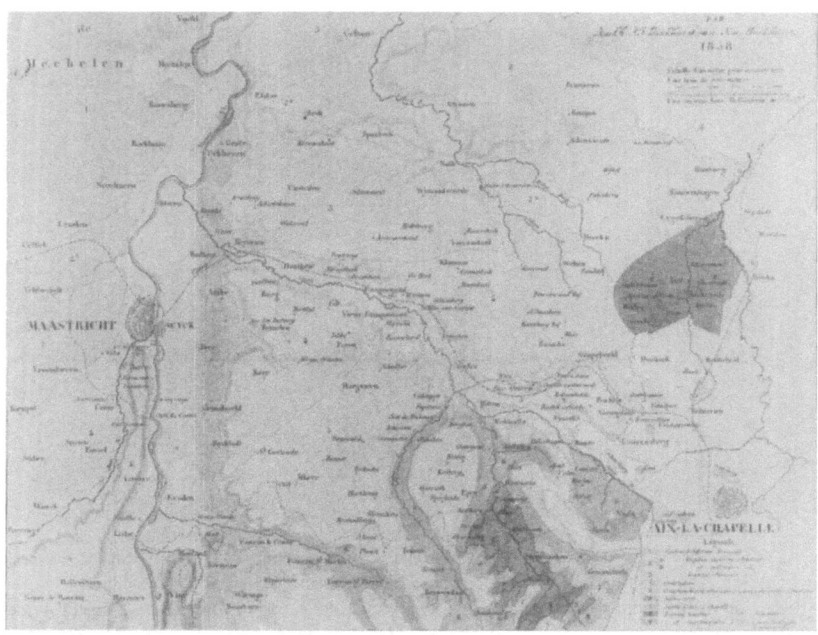

Figure 1.3 - The geological map of South Limburg by Binkhorst (1859). The 50 km wide area forms the northern flank of the Ardennes Massif. The Cretaceous layers between the cities of Maastricht in the West and Aachen in the Southeast dip gently (1°) towards the Northwest. In the highest, southeastern part (321 m) and the river valleys, Carboniferous and oldest Late Cretaceous sediments are exposed, while towards the Northwest increasingly younger Cretaceous-Tertiary sediments are exposed.

In 1799, Le Luc visited the area around Maastricht and noted the resemblance between the sediment of Maastricht and the french "Craie". Both consist almost entirely of calcium carbonate ($CaCO_3$), present as particles that are the broken skeletal remains of marine organisms. Omalius d'Halloy (1822) agreed with the close resemblance between the Chalk of Maastricht and the "Craie", but he also mentioned some differences. While the French Chalk was white and very fine-grained, the carbonates of Maastricht were yellowish, much coarser grained and very porous. Omalius d'Halloy (1822) named the carbonates of Maastricht "Craie Tuffeau de Maestricht". It was named Tuffaceous Chalk because it resembles volcanic tufa. Omalius d'Halloy (1822) placed the Tuffaceous Chalk "officially" within the Cretaceous. The latter was understood to be a geological period and a region, "le terrain Crétacé", in which all Chalk had been deposited.

Fitton, Honey & Connybeare (1829) and Fitton (1834) did not only visit the Tuffaceous Chalk of Maastricht, but they extended their excursion towards the city of Aachen to the East and described a superposition of different sediments that

| TERTIARY | Etage | $-10^6$ y Harland 1982 | Type locality | Author |
|---|---|---|---|---|
| | Danian | 65 - 62 | Stevns Klint, Denmark | Desor 1846 |
| UPPER CRETACEOUS | Maastrichtian | 73 - 65 | Maastricht, the Netherlands | Dumont 1849 |
| | Campanian | 83 - 73 | Aubeterre-sur-Dronne, France | Coquand 1857 |
| | Santonian | 87.5 - 83 | Saintes, France | Coquand 1857 |
| | Coniacian | 88.5 - 87.5 | Cognac, France | Coquand 1857 |
| | Turonian | 91 - 88.5 | Tours, France | D'Orbigny 1842 |
| | Cenomanian | 97.5 - 91 | Le Mans, France | D'Orbigny 1847 |

| | | | |
|---|---|---|---|
| Danian | | Bioclastic sandstone | |
| Maastrichtian | Craie Tuffeau (Marnes de Kunrade) | Bioclastic siltstone | |
| | Craie a Silex | | |
| | Craie Marneuse | Bioclastic mudstone | |
| Campanian | Craie Blanche | Smectitic claystone | |
| | Sables Verts a B. quadrata | Glauconitic quartz sst. | |
| Santonian | Sable D'Aix la Chapelle | Quartz sandstone | |
| Paleozoic | Houlliere | Limestone,Shale,Quartzite & Coal | |

Figure 1.4 - The stratigraphic subdivision of the Late Cretaceous/earliest Tertiary in NW Europe. Estimated ages (Harland et al, 1982, 1989), type localities and authors. The stratigraphy of South Limburg (The Netherlands) after Binkhorst van den Binkhorst (1859).

closely resembles the succession of the English Chalk upon the underlying Sands of Shanklin (Lower Greensand). They discovered that towards the southeast the Tuffaceous Chalk had been removed by erosion and that the deeper and older sediment exposed ("Craie Blanche et Marnes Glauconifère") was much more like the English and French Chalk with and without flint. Even further to the southeast, in the neighbourhood of the village of Vaals, this sediment had also been eroded and an even older sediment cropped out ("Sables Verts"), that consists of quartz ($SiO_2$) sand with a low quantity of carbonate, but with many green particles (glauconite), and therefore was a Greensand.

Davreux (1833) further investigated the succession and discovered sands, near Aachen, that were even older than the Greensand. These were yellowish quartz sands without glauconite that he named "Sables d'Aix en Chappele". He also found a contact between the oldest Cretaceous sediment and the underlying Carboniferous sediment. The Carboniferous formed a sequence consisting of an alternation of sands, clays and coal beds ("terrain houillère" et "anthraxifère").

Dumont (1849), finally, has named the mayor sediment types, based on their typical petrology and fossil content. He depicted the distribution of various systems on the Geological Map of the Belgian Kingdom. He divided the Cretaceous into 5 Systems, which he named "système Achénien, Hervien, Nervien, Senonien and Maestrichtien" respectively. Furthermore he investigated and named the sediments that were younger than the Cretaceous and that covered the "système Maestrichtien" in the North.

Directly upon the Tuffaceous Chalk one could find a yellowish quartz sand, locally greenish due to glauconite admixtures and with few marine fossils ("système Tongrien"). This was covered by sandy clays ("système Rupelien"), and at the top fine-grained white sands with many plant remains, forming lignite layers ("système Bolderien") occurred.

The lithostratigraphic division of the sediments in spatial and temporal units, based on lithology alone, had validity in local areas. If one tried to relate deposits between different areas, separated by large distances, barren of exposures of the sediment of interest, problems arose. For instance, Roemer (1840) did not consider the Tuffaceous Chalk of Maastricht as a separate, youngest system of the Cretaceous, or even as a sediment that represented the period in between the Cretaceous and the Tertiary as had been suggested, but as a true time-equivalent of the English Upper Chalk.

Such different interpretations of the age of the various Cretaceous units, or of the chronostratigraphic position of the different lithological systems, could be clarified using "guide fossils". As fossil collections were completed, it was recognized that particular fossils were restricted to specific types of sediment. Other fossils showed a more general occurrence with respect to lithology. They were restricted, however, in their vertical distribution. These latter fossils were used to date lithostratigraphic units and the concept of biostratigraphy developed.

Pomel (1848), considering the fossil content, argued that the "Craie Blanche" of South Limburg, for instance, was, although lithologically similar, in fact much younger than the Cenomanian and Turonian English Chalk or the French "Craie" along the river Seine. The "Craie Blanche" in the Limburg area should be considered as a local phenomenon, and all Cretaceous sediments of South Limburg were only slightly older than the "Calcaire Pisolithique" near the city of Paris. They belonged to the youngest Cretaceous, the "système Senonien" of d'Orbigny (1840-1842) and formed a Santonian to Maastrichtian micro-sequence of approximately 200 m thickness, that showed the same succession of different facies as had been recognised at different locations in Northwest Europe and Southern France (Triger, 1857).

## 1.2 Litho- and biofacies

According to Hutton (1726-1797), similar rocks had been formed by similar processes that occurred at different times and places, and that could still be observed in modern environments. Gressley (1814-1865) noted that a rock unit with a particular lithology and fossil content had a facies that reflects a set of conditions and processes characteristic for a specific environment of formation. Concerning the deposits that occur in South Limburg, Binkhorst discussed current hypotheses, presumed environments of formation and he distinguished:

### 1.2.1 Formation Houllière

The Cretaceous sediment of South Limburg covers the strongly karstified and abraded surface of the Paleozoic. Paleozoic sediments are shallow marine and continental sands, shales and coal measures that crop out near the village of Vaals and within an isolated patch near the village of Kerkrade. Both areas are separated by the "Feldbiss" fault. West of this fault the Carboniferous is relatively thin and it has been considerably eroded before it was covered by Cretaceous sediments. East of the fault, the Carboniferous is much thicker, the Cretaceous is absent and the Carboniferous sediment is covered by Tertiary sands. Binkhorst recognised that the Paleozoic basement had moved before, during and after the Cretaceous. He concluded that this motion was complex and oscillatory (inversion tectonics): "Il faut donc supposer qu'un soulèvement d'une partie de ce bassin, a été suivi par une dénudation..".

### 1.2.2 Sable D'Aix la Chapelle

The oldest Late Cretaceous, the Sands of Aachen, were studied in detail by Debey (1849, 1857) because of its rich flora. A clay layer with clasts of reworked Paleozoic sediment was found at the basis of the Aachen Sands, indicating the start of the transgression of the Northern flank of the Ardennes Massif by the Cretaceous Sea.

Upwards in the succession the sediments consist of non-consolidated quartz sands. Between the sands, clay layers occur that contain numerous leaves and twigs of marine and continental Cretaceous plants, as well as rare remains of insects. Layers with roots of mangrove trees, *Fucoïdes* and *Nadaïdes* (Debey, 1857), covered by sandy layers with marine fossils and "des corps cylindriques verticeaux de formes variées traversent fréquement les bancs" (trace fossils) reflect the ongoing transgression.

The uppermost part of the Sands of Aachen contains concretions of sand, lithified by silica, that are (sub)horizontally bedded and that very much resemble flint. Locally, silicified trunks of the pine tree *Pinites aquisgranensis* can be found. These have been thoroughly bored by marine pholades and indicate the near coastal environment and the washing ashore of driftwood. The fossilized flora contains the remains of plants that flourished under subtropical climatic conditions, very much like those of the modern "Nouvelle Hollandie" (Debey, 1857).

### 1.2.3 Sables Verts à *Belemnitella quadrata*

The Sands of Aachen are covered by the Sands of Vaals and the boundary between both units is formed by a banded greyish to yellowish conglomeratic sand with grey-white pebbles of the size of a pigeon egg.

The Sands of Vaals consist of yellowish-greenish fine-grained glauconitic quartz sand with less than 10 % carbonate. Moreover the basal deposits of the Green Sands contain pieces of coal and remains of plants. At the middle and top of the several tens of metres thick sequence, several half a metre thick calcareous layers, containing abundant molds of marine fossils, alternate with somewhat thinner layers of quartz sand which are soft and have a low carbonate content. Near Vaals the glauconitic quartz sands locally contain concentrations of silicified fossils and concretions of sand, cemented by silica.

The greenish glauconitic quartz sands are covered by dark bluish-greenish smectitic clays and very fine micaceous sands, that contain the guide-fossil *Gyrolithes*: "le fossil le plus remarquable de ce système est un corps en forme de baguette contournée, que l'on a rapporté à des fucoïdes ou à des Annélides" (Omalius d'Halloy, 1848). The carbonate content of the clayey part of the Sands of Vaals increases gradually upwards in the sequence. The clayey sediments are overlain by a conglomeratic layer with coarse-grained skeletal remains and glauconite. In the neighbourhood of the village of Slenaken, the layer consists almost entirely of the rostrae of *Belemnitella mucronata* (Belemnite graveyard).

Within the green sands and the clays with *Belemnitella quadrata*, Binkhorst van den Binkhorst collected a rich fossil fauna with 3 serpulid species, 16 cephalopod species, 126 gastropod species and 101 pelecypod species. All fossils are remains of animals that were adapted to a soft and mobile sandy/muddy sea bottom or independent of the substratum. Sessile species that required clean water and/or a hard substratum, without repeated pollution by siliciclastic sediment, like e.g. corals, were not found in the fossil association.

The fining upwards and the increase of carbonate content in the succession of the Sands of Vaals reflect the ongoing transgression and the southwards migrating coastline.

### 1.2.4 Craie Blanche, Marnes et Craie à Silex

The calcareous clay or Smectite is covered by pure carbonates. These are, at the basis, represented by a true white Chalk, which can be used to write on the blackboard. In the lowermost part of this Chalk coarse fossil fragments and glauconite grains are abundant. The top of the sequence of soft white Chalk, with few small, black coloured flint nodules and *Belemnitella mucronata* is formed by a well-lithified Chalk.

Above this lithified Chalk, a marly Chalk forms the basis of the sequence of "Chalk with Flint". Towards the top of that sequence, the relatively impure marly

Chalk (80% CaCO$_3$) changes gradually into a pure (97% CaCO$_3$) and somewhat coarser-grained silty Chalk. Also increasingly more and better developed flint layers are present in a regular succession.

About 20 flint nodule layers were recognised by Binkhorst in the uppermost part of the "Chalk with Flint" exposed along the River Meuse. Dumont (1849) noted that these flint nodule layers, separated by, and alternating regularly with Chalk without flint nodules, could be traced along the westbank of the River Meuse over a distance of several kilometres. He also noted that the layers parallel each other, that the thickness and the distance between the layers increases gradually upwards in the series and that in fact all layers dip gently (1°) towards the Northwest.

The origin of the silica concretions had been discussed by Ehrenberg (1812). He noted that the fine-grained carbonates in Sicily contain, contrary to the silica nodule layers common in the Chalk of northern Europe, marl layers with "des infusoires à test siliceux", while the Chalk contained "des Polythalames à test calcaire". Therefore the hypothesis was justified, that the flint in the Chalk originated from the dissolution of siliceous tests and that the dissolved silica had been concentrated subsequently into inorganic formes around different centres of precipitation.

Concerning the fossil content of the Chalk, it was observed by Binkhorst that this was much poorer than in the underlying Smectite and Sands of Vaals. Fossils found were mostly brachiopods, crinoids, irregular echinoids, thin-shelled pelecypods, bryozoans, remains of vertebrates and, above all, remains of very small organisms such as foraminifera and ostracods. All these fossils are calcitic skeletal parts, whereas molds or silicifications of aragonitic skeletons, common in the siliciclastic sediments below, were not found in the Chalk.

According to Nelson, cited in Lyell's (1854) "Manual of elementary Geology", the bottom of the lagoons between the isles of Belize, surrounded by coral reefs, consisted of a white, fine-grained mud, that had been derived from the decomposition of reef debris that, after removal of plant material and drying, could not be distinguished from the ancient Chalk. Godwin Austen (1858) noted that most pelecypods in the Chalk were the remains of animals which attached themselves with a byssus, to e.g. floating algae, and thus would not have necessarily lived on the muddy bottom of the ancient sea. He considered the muddy bottom of fresh Chalk sediment not suitable for most organisms to live on, and considered all fossils as allochthonous, transported towards a tranquil environment after the animals had died.

### 1.2.5 Craie Tuffeau, système Maestrichtien de Dumont

The Tuffaceous Chalk of Maastricht is equivalent to the Maastrichtian s.s. of Dumont (1849) (Fig. 1.5). It is a skeletal carbonate sand of which the type section is exposed at mount St. Pieter near Maastricht. The lower boundary is formed by the "Couche à Coprolites" and the basal part is very similar to the underlying "white Chalk with Flint". However, flint is less common and occurs as curved

Figure 1.5 - Drawings of the lithologic succession of the White Chalk with planar-parallel flint nodule layers (right) and of the Tuffaceous Chalk (left) with lenticular fossil grit layers at the floor of the caves, isolated flint nodules in the sediment above and a hardground at the ceiling of the caves, covered by bryozoan-rich fossil grit layers (Binkhorst van den Binkhorst, 1859).

laterally restricted nodule layers, instead of (sub)horizontal and laterally continuous layers.

The upper part of the Tuffaceous Chalk was excavated for building purposes. This Chalk is a medium- to coarse-grained glauconitic bioclastic sand. It contains lithified layers which have been encrusted by bryozoans, serpulids and oysters and which have been bored by lithophages.

Reworked and encrusted lumps of lithified Tuffaceous Chalk cover the lithified layers locally and have been admixed with a coarse-grained skeletal sand, very rich in the remains of bryozoans, large foraminifera, algae and the molds of solitary and colonial corals ("Couches à Bryozoaires"). Laterally the Tuffaceous Chalk changes into a less pure carbonate, the "Marl of Kunrade", that was considered by Binkhorst to represent a true lateral time-equivalent of the Tuffaceous Chalk of Maastricht. The Marl of Kunrade contains pieces of coal and the remains of marine plants, e.g., *Thalassocharis bosqueti* and land plants, e.g., *Sequoia cryptomerioides*. Therefore it was thought to represent the near-coastal facies of the Tuffaceous Chalk, deposited in the vicinity of a river mouth.

The Maastrichtian Tuffaceous Chalk has a very rich and diverse fossil association. Binkhorst recognised 241 genera and 798 species of which the bryozoa with 60 genera and 280 species were most abundant, followed by pelecypods (35 genera, 115 species), crustacea (30, 89), gasteropods (20, 80), foraminifera (18, 39), anthozoans (15, 33), fish (14, 28), echinoids (13, 42), brachiopods (10, 35), cephalopods (8, 17), annelids (5, 17), sponges (5, 13), crinoids (4, 5) and reptiles (4, 5).

The fossils in the Tuffaceous Chalk and the Marl of Kunrade, and the coarse grain size of these sediments, led Binkhorst to believe that the facies was comparable with the facies of the near-coastal Sands of Vaals, and that the Tuffaceous Chalk of Maastricht and the Marl of Kunrade represent the regression of the Late Cretaceous Sea.

### 1.2.6 Pierres Cornées et Orgues Géologiques

The Cretaceous carbonate sediment suffered from dissolution after its deposition. According to Beissel (cit. Binkhorst, 1859), the low carbonate content in the silici-clastic sediments of Aachen, Vaals and Kunrade was even the result of syn-depositional dissolution of carbonate by means of acid meteoric water. Dumont (cit. Binkhorst, 1859) considered the lower carbonate content to be the result of upwelling of acid hydrothermal fluids along faults. The area was known for its thermal bath.

However, most of the dissolution was probably a late Tertiary phenomenon. In the higher parts of South Limburg which have not been affected by erosion by Pleistocene rivers, one finds locally "Pierres Cornées". These are irregular flints, slightly rounded by dissolution, but not by transport. They occur in a reddish clay, that covers the Green Sands and the white Chalk with Flint. This sediment is a residue, that remained after the carbonate of the Chalk with Flint had been dissolved. These residual deposits witness the importance of late post-depositional dissolution by meteoric water.

Interesting is the report by Dumont (1847) of such deposits exposed at Hautes Fagnes, about 25 km south of Vaals, at the highest point of the Ardennes Massif, 680 metres above the present sea level. The dissolution residue with fossils in flint indicates the former presence of White Chalk with Flint. The flint nodules are superimposed in their proper stratigraphic succession and the eluvium reflects the post-Cretaceous uplift of the Ardennes Massif. As Dumont already noted: "elle est nécessairement le résultat d'un movement postérieur à la formation crétacée; car on ne peut admettre que les espèces d'êtres organisés dont on trouve les restes à Francorchamps, aient pu vivre, en même temps, dans ces lieux, vers la surface, et à Maastricht, sous la pression énorme d'une colonne d'eau de 600 mètres."

Other features, which are a result of postdepositional karst, are the "Orgues Géologiques" (Mathieu, 1813). These are several metre wide cavities, that start in the uppermost part of the Tuffaceous Chalk and that may descend more or less vertically, towards a depth of more than 30 metres into the white Chalk with Flint below. The cavities are rarely empty and have been mostly filled with clay, sand and gravel derived from the Tertiary deposits covering the Tuffaceous Chalk. According to Buckland (1839), these "Sand Pipes" were also the result of continuous infiltration of the limestone by rain water, containing carbonic acid, dissolving the limestone, and allowing the overlying sediment to slide down into the thus produced cavities.

# Conclusion

The Late Cretaceous sediments of South Limburg have not only shaped the characteristic country side of gentle hills with a lush vegetation, but they also influenced the history and culture of the people that inhabited the area since thousands of years.

The geology of South Limburg was subject of discussion since the Renaissance and the start of the science of geology in Europe. Halfway the last century this had already led to an excellent understanding of the geological history of the area. The observations, interpretations and conclusions of the above quoted authors are considered to be still valid.

Summarizing, they are: The abraded and karstified surface of the Palaeozoic basement was flooded by a subtropical sea during the Senonian. The basement was tectonically active and the motion continued during the deposition of a Late Cretaceous transgressive-regressive sequence. The transgressive part at the basis consists of siliciclastic deposits, the lagoonal Sands of Aachen and the shallow marine glauconitic sands of Vaals. The regression is represented by shallow marine carbonate sands, the Tuffaceous Chalk of Maastricht and by the Marl of Kunrade. The sediments that occur between the coarse-grained transgressive and regressive facies are fine-grained mudstones like the Marly Chalk, the White Chalk and the White Chalk with flint. The Late Cretaceous deposits were uplifted above sea-level during the Tertiary, and they suffered considerably from dissolution by acid rain water intrusion and karstification.

# 2 Present-day Views on Chalk

## Introduction

In the 19th century it was already recognised (Binkhorst van den Binkhorst, 1859) that the 200 m thick Campanian-Maastrichtian Greensand, Smectite, Chalk and Tuffaceous Chalk succession of South Limburg (The Netherlands) is a transgressive-regressive microsequence, in analogy to the general facies succession of the Late Cretaceous of NW Europe.

During the last 150 years, the knowledge of the environment of Chalk deposition has steadily increased and some aspects of the (Tuffaceous) Chalk of South

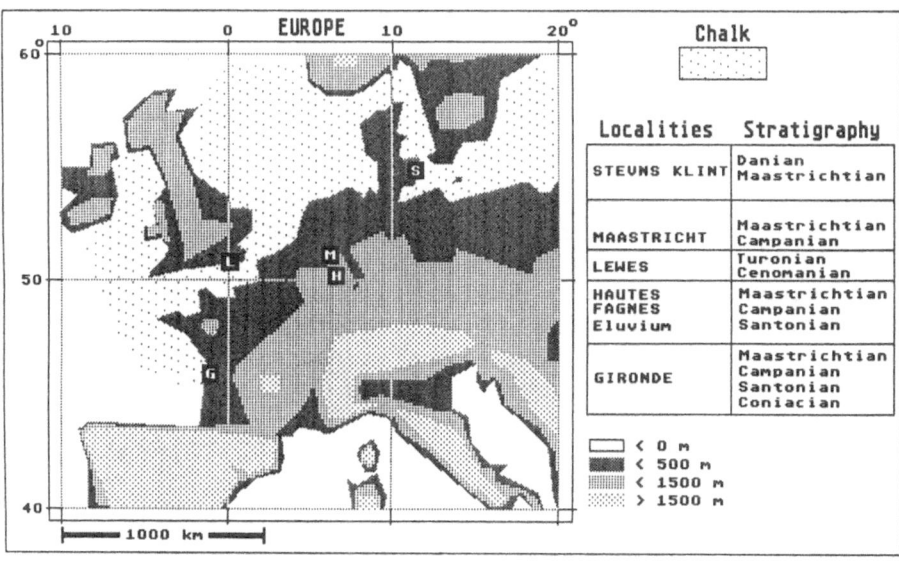

Figure 2.1 - The present topography of Europe and the distribution of Chalk, preserved below 200 m (pre-Glacial sea level). At Hautes Fagnes a dissolution residue (eluvium) occurs at 650 m above present-day sea level. Areas that will be discussed in this thesis have been marked.

Figure 2.2 - SEM picture of pure Chalk of the Zeven Wegen Member (Gulpen Formation, Late Campanian, quarry North, Lixhe, Belgium) consisting mainly of micron-sized debris of coccolithophores.

Limburg in particular, and of the Chalk of NW Europe are summarized, based on part of the more recent literature on Chalk and related sediments.

Chalk is a light-coloured, fine-grained, porous and friable carbonate mud rock that is probably best known as blackboard chalk and from the "white cliffs of Dover". The Chalk was deposited in NW Europe during the Late Cretaceous and it has been preserved in several basins north of the Alpine Mountain Belt (Fig. 2.1). The stratigraphic succession consists of 7 Stages (Fig. 1.4). The youngest Stage (Danian) lacks fossils such as ammonites, belemnites, rudists and dinosaurs and therefore belongs to the Tertiary. However, the Danian is closely related to the Late Cretaceous because it is also represented by Chalk.

The Chalk consists for a larger part of very fine-grained remains of calcitic skeletons of marine organisms and therefore resembles pelagic carbonate oozes of the present deep sea (cf. Schlanger & Douglas, 1974). Like in deep-sea oozes, the particles of Chalk fall within the suspension fraction (microns) and coccoliths, micron-sized, low Mg-calcite remains of haptophycean uni-cellular algae, the coccophores are most abundant (Paasche, 1968; Fig. 2.2).

In recent deep-sea ooze, coccoliths are frequently mixed with micron-sized remains of other pelagic uni-cellular plants such as diatoms with siliceous tests (Calvert, 1974) or they are mixed with somewhat larger mm-sized unicellular planktonic animals such as amoeboid foraminifera with a calcitic test (Schlanger & Douglas, 1974) and radiolaria with a siliceous test (Garrison, 1974). In the Late Cretaceous Chalk, the siliceous skeletons of radiolaria and diatoms have been rarely preserved (Soudry et al., 1981), supposedly because they dissolved during late diagenesis while the dissolved silica precipitated as flint (Ehrenberg, 1812).

There are more differences between the Cretaceous Chalk and recent deep-sea oozes. For instance, remains of benthic organisms are much more common in the Chalk than in recent deep-sea ooze. Furthermore, Cenozoic deep-sea coccolith ooze is found alternating with other deep-sea sediments (e.g. radiolarites), whereas the Late Cretaceous Chalk of NW Europe is resting on, and covered by, shallow marine and continental terrigenous sediments. Therefore, the Late Cretaceous Chalk of NW Europe is considered to have been deposited on the floor of a comparatively shallow epi-continental sea (Håkansson et al., 1974), that has no recent equivalents.


## 2.1 Particle composition of Chalk


In Maastrichtian Chalk, the <0.063 mm grain-size fraction typically forms 90% of the solid rock volume (Bromley, 1979). Non-carbonate particles form less than 1% of the rock volume in the purest carbonates. Insoluble particles are mixed-layer clays such as montmorillonite/illite (Christensen et al., 1973; Thorez & Monjoie, 1972) and very fine-grained quartz dust (van Harten, 1972; Håkansson et al., 1974). At high concentrations of fine-grained insolubles, the sediment is marly (smectitic) Chalk.

The <0.063 mm carbonate fraction consists mainly (>60%) of coccoliths and the rest of this fraction is debris of planktonic foraminifera, calcispheres and fine debris of skeletons of larger benthonic/nectonic organisms. Skeletal remains of larger benthonic/nectonic organisms in older Chalk are typically the remains of inoceramids. In younger Chalk bryozoans generally form the coarser fraction (Bromley, 1979).

Furthermore, the skeletal parts of sponges, brachiopods, octocorals, echinoderms, benthic foraminifera, and serpulids are common in the >0.063 mm grain-size fraction (Felder et al, 1985). Large skeletal remains are rare in the fine-grained Chalk and their diversity is low. The fossils are, however, well preserved and hardly abraded. In coarser-grained Chalk, the coccolithic ooze may be mixed with quartz sand, glauconite and/or phosphate and is named Sandy Chalk, Glauconitic Chalk (Greensand) and/or Phosphatic Chalk respectively. Where silt-sized and

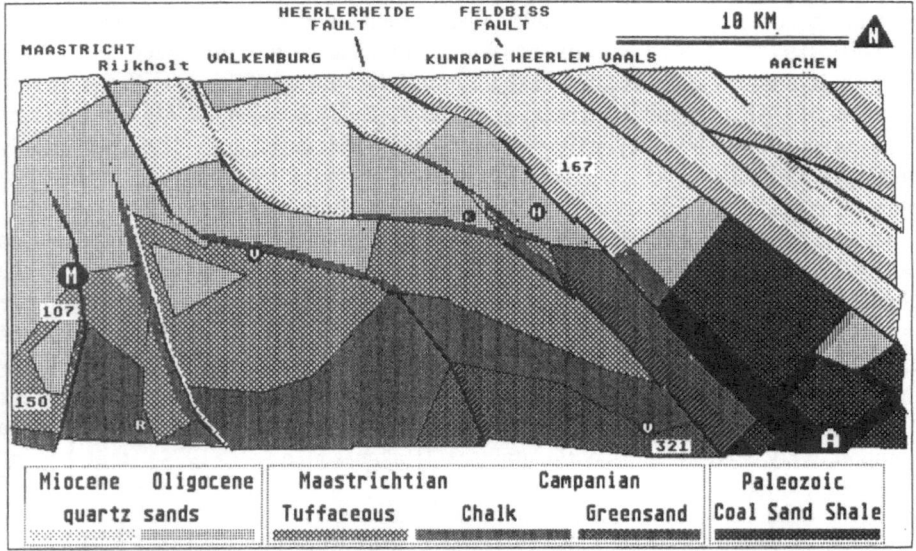

Figure 2.3 - The block-faulted, NW-dipping flank of the Ardennes Massif in South Limburg. Blocks of karstified, abraded Paleozoic sediment descend stepwise towards the NE (direction Rhine Graben). The NW dipping fault blocks are covered by Cretaceous (Campanian-Maastrichtian) shallow marine siliciclastics/carbonates and by Tertiary (Oligocene-Miocene) shallow marine-continental siliciclastics. The sediments were eroded and karstified during the Pleistocene (after Felder & Bosch, 1984).

sand-sized bioclasts are the dominant particles, the sediment is named Tuffaceous Chalk (Omalius d'Halloy, 1822). In the coarse-grained facies, large skeletal remains are abundant and diverse. They are concentrated in layers and lenses, and are frequently less well preserved, showing signs of abrasion.

## 2.2 Paleogeography

Maastrichtian Chalk is thought to have been deposited in a seaway running from the Atlantic in the west, through the North Sea, into Poland to the east, where the marine basin widened into the Russian shelf sea further to the east (Håkansson et al., 1974; Ziegler, 1982). Maastrichtian Tuffaceous Chalk also occurs in SW France (Séronie Vivien, 1972). It is doubtful whether the present distribution of Maastrichtian Chalk deposits reflects the full extension of the Late Cretaceous Sea over the North West Eurasian continent (Voigt, 1929). For instance, an isolated

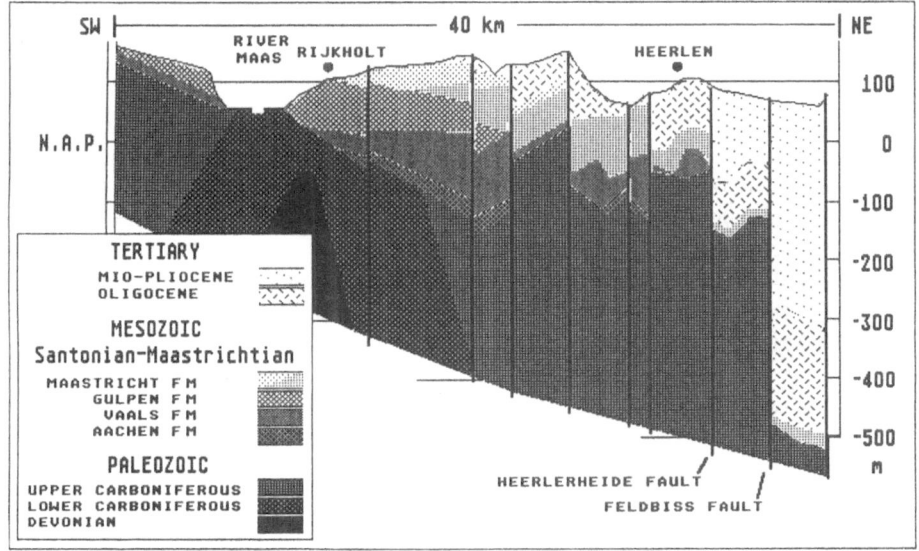

Figure 2.4 - SW-NE cross-section through the block-faulted flank of the Ardennes Massif. The Paleozoic subsurface with the Visé-Puth antiform (overthrust) and faults with unknown dip. The older Paleozoic is covered by Cretaceous in the SW and the younger Paleozoic is covered by Tertiary in the NE (after Felder & Bosch, 1984).

patch of Maastrichtian flint eluvium occurs on the 700 m high Ardennes Massif (Hautes Fagnes, Belgium), which indicates post-depositional uplift, dissolution and erosion of Late Cretaceous Chalk on at least this present-day high (Dumont, 1847; Renier, 1902; Legrand, 1968; Gullentops, 1987).

Near-coastal deposits that may be used to identify the outline of the Maastrichtian Sea are either rare or have not been recognised as such. The biomicritic Chalk passes laterally and vertically downwards into Marly Chalk and Greensand, which cover a continental basement. Towards the top of the Chalk, a coarse-grained Tuffaceous Chalk has been locally preserved. Both coarse-grained facies have been interpreted as being basin-marginal, reflecting near-shore deposition (Binkhorst van den Binkhorst, 1859; Voigt, 1929).

## 2.3 Sequence thickness and rates of deposition

The thickness of the Maastrichtian Chalk varies. In the subsurface of the North Sea a thickness of 200-400 m is common. Locally (Danish Basin) it may attain a thickness of 700 m (Heybroek et al., 1967; Stenestad, 1972). Towards the present continental massifs the Maastrichtian Chalk succession becomes thinner and it is practically absent above an altitude of 200 m, roughly corresponding with the pre-glacial sea level (Guilcher, 1969).

The duration of the Maastrichtian is estimated at 8 million years (65-73 Ma; Harland et al, 1982, 1989) and given the thickness of the Maastrichtian Chalk sequences (700-200 m), a mean rate of deposition between 10 cm/ka in the central North-Sea basin and 2 cm/ka at the margins of the massifs seems plausible. Such values are comparable with deposition rates of 4 cm/ka for the Maastrichtian (van Hinte, 1976), 4.5 cm/ka for Cenomanian-Campanian Chalk of England (Black, 1953) and 18-30 cm/ka for the Maastrichtian Chalk of Rügen (Germany) (Nestler, 1965; Håkansson et al., 1974).

## 2.4 Sea level, tectonics and depth of deposition

The depth and the width of the Maastrichtian marine basin changed as the result of global sea level fluctuations (eustasy; Suess, 1900), motion of the Earth crust (isostasy; Haug, 1900) and/or simultaneous changes of sediment production, transport, deposition and erosion.

Assuming that the depth of deposition hardly varied, the rise of the sea-level, relative to the basement, should have been proportional to the thickness of the Late Cretaceous sequence. A Late Cretaceous sea-level rise of 200-450 m has thus been proposed (Bond, 1978; Robaszynski, 1981).

Considering the height difference between the Cenomanian at present sea level (e.g. Beachy Head, England) and the Maastrichtian at 650 m above sealevel, in the Ardennes Massif, and assuming that the basement of the London-Brabant-Ardennes Massif was stable during and since the Cretaceous, a Late Cretaceous sea-level rise of 650 m was proposed (Hancock & Kauffman 1979).

However, the assumed stability of the basement is doubted. A facies and thickness change of the Late Cretaceous deposits, in a direction oriented perpendicular to large recent faults, is considered evidence for different motions of fault-bounded blocks during the Late Cretaceous (Binkhorst, 1859; Breddin, 1932; Legrand, 1961; Robaszynski, 1981; Bless, 1988) (Figs. 2.3, 2.4)

It is thought that during the Cretaceous, the vertical motions of fault blocks were opposite to the motion of these blocks before and after the Cretaceous (inversion, Breddin, 1929; Muller, 1945; Patijn, 1961). The motion along the faults is thought to have been complex (cake walk), mostly reversed, locally normal and with a rate of 0.1-0.2 cm/ka during the Late Cretaceous (Rossa, 1988).

On the other hand, it was argued that the fault blocks have been displaced laterally during the Tertiary (Umbgrove, 1926), the difference between facies and

| Ma | Series | Stage | | Formation | Member | | |
|---|---|---|---|---|---|---|---|
| -65 | Lower Tertiary | Danian | | Houthem | Geleen | | R |
| | | | | | Bunde | Md | Q |
| | | | | | Geulhem | | P |
| | | | | | Meerssen | | NML |
| | | LATE | | Maastricht | Nekum | Mc | K |
| | | | | | Emael | | I |
| | | | | | Schiepersberg | Mb | H |
| | | Maastrichtian | | | Gronsveld | | |
| | | | | | Valkenburg | Ma | G |
| | | | | Gulpen | Lanaye | Cr4 | F |
| | | | | | Lixhe 3 | | |
| | | LATE | | | Lixhe 2 | Cr3c | E |
| | | | | | Lixhe 1 | | |
| | | | | | Uylen | Cr3b | D C |
| -73 | Upper Cretaceous | EARLY | | | Beutenaken | | B |
| | | LATE | | | Zeven Wegen | Cr3 | A |
| | | Campanian | | Vaals | Terstraeten | | |
| | | | | | Beusdaal | | |
| | | EARLY | | | Vaalsbroek | | |
| | | | | | Gemmenich | Cr2 | A' |
| | | | | | Cottessen | | |
| | | | | | Raren | | |
| -83 | | | | | Hauset | | |
| | | Santonian | | Aken | Aken | Cr1 | Foram zones |
| | | | | | Hergenrath | | |
| | | Coniacian | | Formations | Members | | Hofker 1966 |
| | | Turonian | | | | | |
| | | Cenomanian | | | Felder 1975ab | Uhlenbroek 1912 | |

Figure 2.5 - The lithostratigraphy (Uhlenbroek, 1912; Felder, 1975a,b) and benthic foraminiferal bio-stratigraphy (Hofker, 1966) of the Late Cretaceous of South Limburg.

**Aken Formation:** At the basis lagoonal clays of Hergenrath cover the karstified Paleozoic sediments (Breddin et al, 1963). These are covered by Santonian (Batten et al., 1988) beach-barrier quartzose Sands of Aachen (Davreux, 1833).

**Vaals Formation:** Early Campanian (Albers, 1976) marine glauconitic quartz sands and smectitic clays (Herve Facies).

**Gulpen Formation:** Late Campanian white chalk, Early-Middle Maastrichtian grey-marly Chalk and Late Maastrichtian white Chalk with flint (Dumont, 1832; Hofker, 1966; Jagt, 1988).

**Maastricht Formation:** Late Maastrichtian carbonate silts and sands (Tuffaceous Chalk) which form the Maastrichtian s.s. of Dumont (1849). Towards the east the Tuffaceous Chalk of Maastricht grades into a marly Chalk (Kunrade Facies) (Felder, 1975a,b).

**Houthem Formation:** Early Tertiary (Dano-Montian) Tuffaceous Chalk (Hofker, 1955, 1956, 1957; Meijer, 1959; Wienberg Rasmussen, 1965; Felder, 1975 a,b).

thickness of the Cretaceous sediment at either side of faults might thus reflect a lateral displacement of deposits of different character and thickness, that were initially deposited far apart and that later moved adjacent to each other.

The complex relation between sea-level variations and vertical motions hampers the definition of the water depth of the Cretaceous Chalk basin. Relatively deep deposition of the hemi-pelagic Chalk below the photic zone and below storm-wave base was proposed, based on: the absence of remains of light-dependent benthic organisms in most of the Chalk (>50 m; Bromley, 1965); the depth ranges of recent muddy bottom communities that are comparable to those found in the

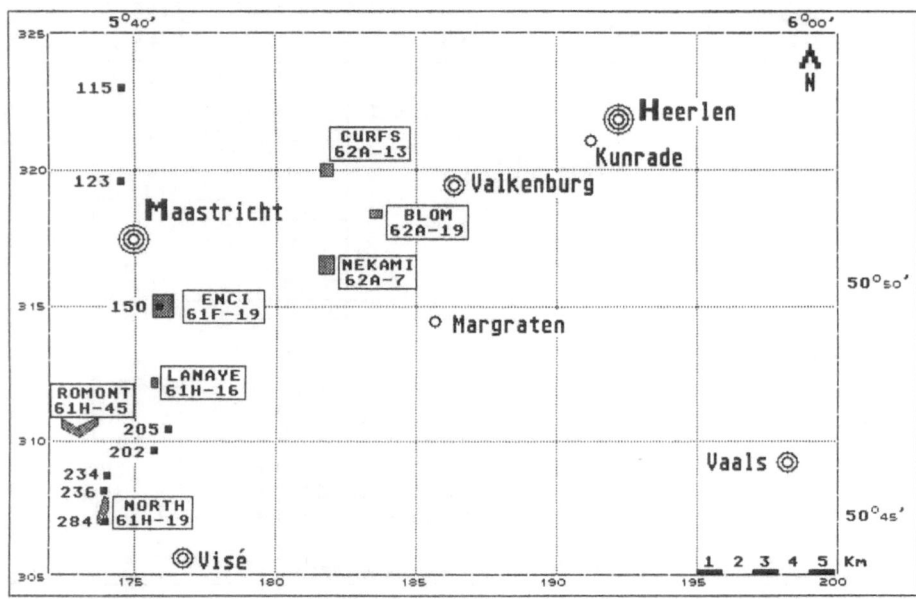

Figure 2.6 - Outcrop of the Campanian-Maastrichtian-Danian (Tuffaceous) Chalk of South Limburg in quarries (numbers refer to Felder, 1975 a,b,); (Fig. 2.7) and in several boreholes along the River Meuse (Figure 2.8).

Chalk (>100-250 m; Nestler, 1965; Reid, 1968) and the absence of aragonite (Hudson, 1967), presumably because Chalk was deposited below the aragonite compensation depth (>60 m; Hancock, 1963). However, there are also indications for very shallow depositional conditions and possibly even emergence of the Chalk during the Maastrichtian. Molds of the gastropod Cerithium have been found in lithified Chalk (Bromley, 1975, 1979). Today Cerithium occurs on restricted (inter-tidal) mudflats (Bathurst, 1971). Probably intertidal algal stromatolites were reported from the Maastrichtian Chalk of Ireland (Bathurst, 1971, p. 405). Other indications for the deposition of Chalk at relatively shallow depth are given by remains of light-dependent organisms in particular in the Late Maastrichtian Tuffaceous Chalk of Maastricht, i.e., hermatypic colonial corals (Umbgrove, 1925), silicified sea grasses (Voigt & Domke, 1955), imprints of plants on the attachment surface of epiphytic organisms and the remains of boring algae (Swinchatt, 1965) and calcareous benthic algae in Tuffaceous Chalk (Voigt, 1929; Umbgrove 1927). Finally, shallow deposition is also suggested by (sub)recent Holocene greensands that are covered by coccolithic oozes, deposited in lagoonal environment in Southern Belize (Simien, 1987).

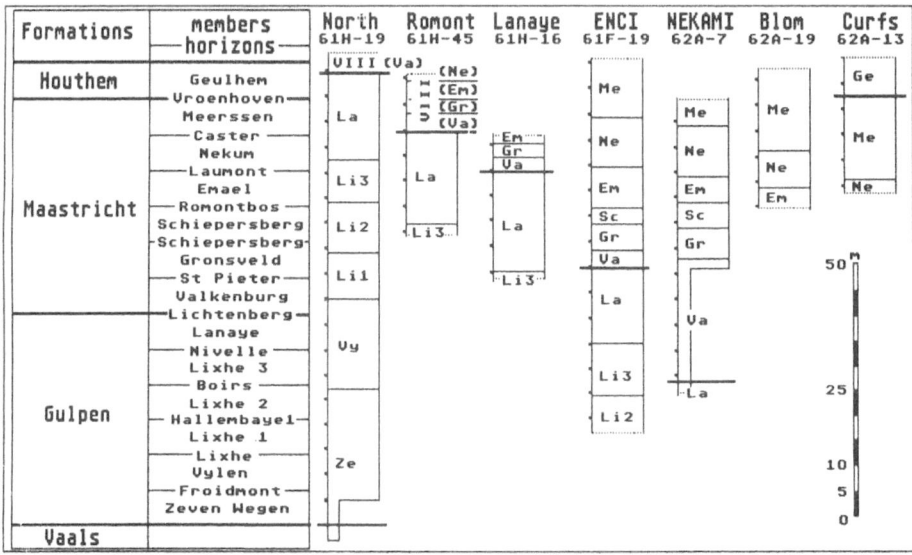

Figure 2.7 - Lithostratigraphic subdivision of the (Tuffaceous) Chalk of South Limburg exposed in quarries (Felder, 1975 a,b,); (Fig. 2.6).

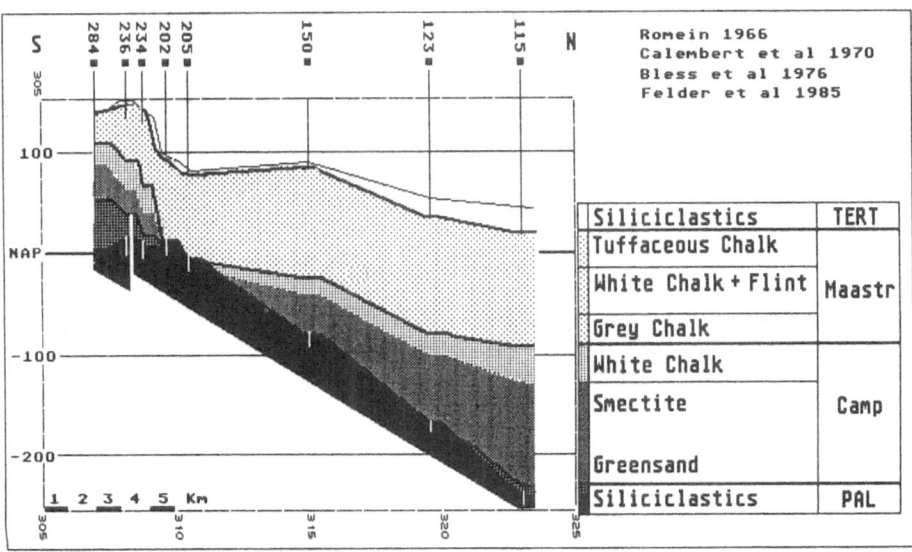

Figure 2.8 - Litho-biostratigraphic subdivision of the Late Cretaceous exposed in boreholes along the River Meuse (Fig. 2.6).

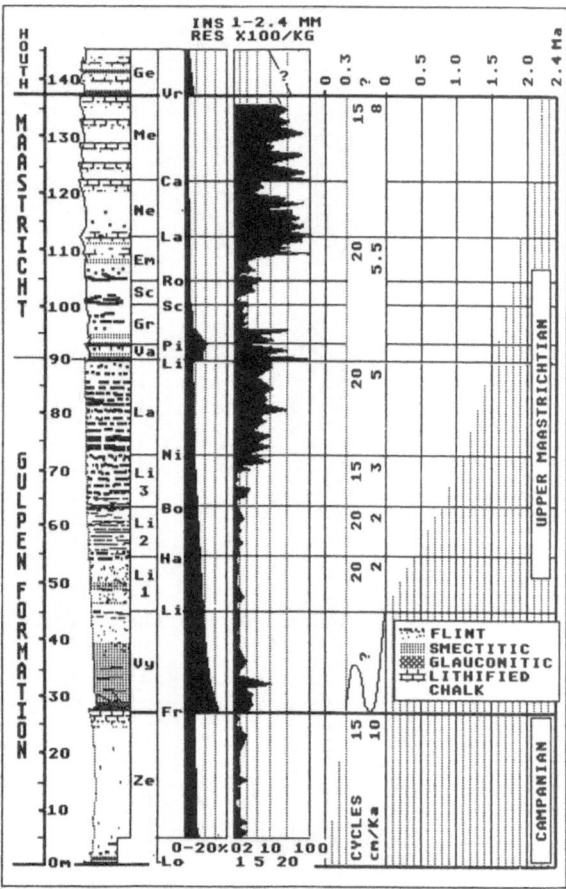

Figure 2.9 - Composite lithostratigraphic column of the Gulpen Formation (Late Campanian and Middle to Late Maastrichtian, quarry North), Maastricht Formation (Late Maastrichtian, quarry ENCI) and Houthem Formation (Danian, quarry ENCI and quarry Curfs). The Maastrichtian s.s. of Dumont (1849) is equivalent with the Maastricht and Houthem Formations. Left of the 150 m thick sequence is depicted the lithification (theoretical weathering profile). Further to the right are depicted: the lithology (see box); Members and Horizons (Felder, 1975a,b); insoluble residue (Villain, 1977); number of 1-2.4 mm large bioclasts (times 100=0-10,000) per kilogram sample (Felder, 1988); number of (precession) cycles counted per interval and estimated mean deposition rate [interval thickness (cm) / (number of cycles x 20 ka)] and; the accretion of the sediment surface during 0.3 and 2.4 million years.

## 2.5 Stratigraphy

Attempts have been made to divide the Maastrichtian Chalk into shorter stratigraphic intervals. A supra-regional stratigraphic subdivision of the Maastrichtian Chalk based on belemnites was proposed by Jeletzky (1951), Schmid (1959) and Christensen (1988). Belemnites in the Late Campanian and Maastrichtian Chalk of South Limburg allow to fit this regional succession into the supra-regional framework (Fig. 2.5). A biostratigraphic sub-division based on brachiopods has enabled detailed correlations on a regional scale (Steinich, 1965; Surlyk, 1970, 1972). Lithostratigraphic correlation on a regional scale also can be performed on the basis of laterally continuous marker horizons such as prominent marl beds, silica concretion layers (flint horizons), lithified layers (hardgrounds), glauconitic fossil-grit layers and strongly or typically bioturbated sediment layers (burrow horizons). Such lithostratigraphic marker horizons have been found to be continuous over distances of several tens of kilometres (Calembert, 1953, 1956; Felder, 1975a,b) or even hundred kilometres and more (Wood & Smith, 1978; Bromley & Gale, 1982).

The thicknesses of the formations and members vary gradually and low-angle unconformities between successive Chalk members are repeatedly found (Romein, 1962, 1963). They are due to condensation and erosion, and they are characterised by mineralized and/or overgrown hardgrounds, that are covered by coarse-grained, winnowed and condensed sediment (Voigt, 1929; Bromley, 1968).

The lithostratigraphic subdivision of the Late Cretaceous of South Limburg (Binkhorst, 1859; Uhlenbroek, 1912; Felder, 1975a,b) is based on the recognition of laterally continuous flint layers and erosion surfaces overlain by coarse, glauconitic skeletal sand. The lithostratigraphic members contain specific assemblages of benthic foraminifera, that allow a correlation of biostratigraphy and lithostratigraphy (Hofker, 1966). Recently, a stratigraphic investigation of the number and type of mm large bioclasts in the Chalk of South Limburg provided new "eco-stratigraphic" marker horizons (Felder 1981, 1982, 1988; Felder et al. 1985) (Figs. 2.6, 2.7, 2.8, 2.9).

Gamma-ray measurements performed on exposures of the Chalk of South Limburg (Felder & Boonen, 1988) have demonstrated that smectite, glauconite and lithification in this Chalk can be recognised and that it can be used to identify lithostratigraphic units in the subsurface. Radiometric measurements on glauconites have been performed (Priem et al, 1975) and a regular pattern in the succession of dm-m thick Chalk beds has been considered an expression of Milankovitch periodicity and may allow an accurate chronostratigraphic subdivision of the Chalk (Hart, 1987; Gale, 1989; Leary et al., 1989; Herrington et al., 1991).

## 2.6 Paleoecology

Within the Maastrichtian Chalk of NW Europe various fossil communities have been investigated and their paleoecological meaning was discussed. In the last century, already 1073 species had been reported from the Cretaceous of South Limburg (Ubachs, 1879). A number of studies have appeared since, dealing with the taxonomy and to a lesser extent the stratigraphy and paleoecology of the various genera found in the Late Cretaceous of South Limburg (Fig. 2.10).

In fine-grained Chalk, fossils are rare and have a low diversity, while fossils are common and very diverse in the coarser-grained Tuffaceous Chalk. The fossils in Chalk are small and/or light-weighted remains of light-independent, suspension and/or deposit feeders, which were adapted to a soft, muddy substratum (e.g. irregular burrowing echinoids, long-stemmed crinoids, thin-shelled oysters and inoceramids).

Barnacles, crustaceans and brachiopodes (Bosquet, 1847, 1854, 1859)
Crustaceans (Binkhorst, 1857)
Rudists (Bayle, 1857/1858)
Gastropods and cephalopods (Binkhorst, 1861; Kaunhowen, 1897)
Ammonites (Grossouvre, 1908)
Bryozoans (Ubachs, 1858)
Echinoids (Lambert, 1911; Meyer, 1965)
Corals and algae (Umbgrove, 1925, 1927)
Fish (Leriche, 1927)
Crinoids (Smiser, 1935)
Ostracods (Van Veen, 1928, 1932, 1934, 1935-36, 1936; Deroo, 1966)
Serpulids (Jäger, 1988)
Sharks and rays (Van De Geijn, 1937)
Rudists (Van De Geijn, 1940)
Seagrasses (Voigt & Domke, 1955)
Brachiopodes (Kruytzer & Meijer, 1958; Kruytzer, 1969)
Sauriens (Kruytzer, 1964)
Foraminefera (Hofker, 1955, 1956, 1957, 1966; Meessen, 1977; Villain, 1977)
Palynology (Vanguestaine, 1966)
Coccolithophoridae (Vangerow & Schloemer, 1967)
Calcisphaerulidae (Villain, 1975).
Paleoecology (Voigt, 1959, 1974; Bless, 1988; Jagt, 1988; Robaszynski, 1988; and Bless & Robaszynski, 1988)

Figure 2.10 - Some publications on the taxonomy, stratigraphy and paleoecology of fossils found in the Campanian, Maastrichtian and Danian of South Limburg.

The fossils in Tuffaceous Chalk are moreover larger, heavier and less buoyant remains of sessile, partly light-dependent suspension feeders, which were adapted to a mobile sandy or rocky substratum (e.g. regular echinoids, asteroids, sea grasses, corals, rudists, large thick-shelled oysters, gastropods and pelecypods). Biohermal facies of bryozoans and corals (bafflestones) are rare and have been observed in the Tuffaceous Maastrichtian-Danian Chalk of Denmark (Thomsen, 1976; Floris, 1979).

The nature of the substratum and of the depositional to early diagenetic conditions of the (Tuffaceous) Chalk are well reflected by trace fossils, the result of bioturbation and bioerosion on all scales (Kennedy, 1970; Ekdale & Bromley, 1991). The tiered ichnofabric still occurs in situ and reflects the layered environmental conditions in seawater and below the sediment surface, i.e., salinity, oxygenation, sedimentation rate, hydrodynamics, bottom strength and bacterial metabolism during decomposition of organic matter (Bottjer & Ausich, 1986; Bromley et al., 1975; Bromley & Ekdale, 1986). The Chalk has been thoroughly bioturbated during oxygenated conditions. The sediment was a soft mud, at shallow depth mixed by unidentified deposit feeders and penetrated to a depth up to several m by deposit feeders, that left spreiten burrows (*Planolites, Taenidium, Zoophycus*), stationary burrows (*Chondrites, Bathichnus*) and open burrow networks in somewhat more consolidated sediment (*Thallassinoides*) (Ekdale & Bromley, 1991).

Locally the soft sediment consolidated into a firm bottom or it was lithified into a hardground (Voigt, 1929, 1959, 1974). In that case, the ichnofossils of burrowing organisms were replaced by those of boring and bioeroding organisms (Bromley, 1968, 1975; Kennedy & Garrisson, 1975; Marshall & Ashton, 1980).

## 2.7 Diagenesis

During early diagenesis, authigenesis of minerals such as phosphate, smectite, glauconite, pyrite, carbonate cement and silica occurred at or close to the sediment surface (Håkansson et al, 1974; Bromley et al, 1975; Clayton, 1986). During late diagenesis and burial, Chalk consolidated due to compaction, dewatering and dissolution/precipitation. Chalk at a considerable burial depth was studied at DSDP site 167 (Central Pacific Magellan rise, Winterer et al., 1973; Schlanger & Douglas, 1974). Maastrichtian Chalk at 700 m burial depth was still more porous (50%) than Chalk in land sections (35-40%, Bromley, 1965; cit. Bathurst, 1971). The high porosity and relatively low degree of cementation at greater burial depth, was attributed to the fine grainsize of the Chalk and the presence of crystallisation-inhibiting magnesium ions (Neugebauer, 1974; Scholle, 1974). Hardened and tightly cemented Chalk formed in Ireland during late thermal metamorphism due to basalt intrusions and heating, and porosity decreased to no more than 5-6.5% (Wolfe, 1968).

During latest diagenesis, emerged Chalk exposed on the land and situated above the groundwater table, was no longer immersed in its pore fluid and compacted. The Chalk oxidized and dissolved as a result of percolating acid meteoric water (Karst; Albers & Felder, 1979).

# Discussion

Studies of the Late Cretaceous Chalk of NW Europe focused attention on various aspects of the environment of deposition and on diagenesis. Students of geology that advocate actualism and cherish the motto "The present is the key to the past" might like to consider the Chalk as an equivalent of recent deep-sea marine sediments, to a large extent preserved in basins that still reflect the Cretaceous paleogeography, which was presumably similar to that of the present NW European continent. On the other hand, one may favour a motto that stresses the importance of geologic research, "The past is the key to the present", and that invites the student of geology to consider the environment of Chalk deposition to be a very different and unique precursor of the present-day marine environments.

The Late Cretaceous Chalk consists of a succession of (sub)horizontal, laterally continuous, dm-m thick beds. Given the small grain size of the terrigenous silici-clastics in the Greensand and the Smectites at the transgressive basis of the Chalk sequences, and given the low concentration of terrigenous siliciclastics in the (Tuffaceous) Chalk, the NW European continent must have had a low relief and a larger part must have been flooded during the Late Cretaceous. There are some arguments to revive the 19th century hypothesis of a very shallow marine depositional environment for Chalk (Binkhorst, 1859).

During the Tertiary, NW Europe moved from a subtropical into a temperate climate zone, i.e., about 1000 km and 10° towards the North (Wegener, 1915; McElhinny, 1973; Barron et al. 1981). During the Tertiary, the Chalk was uplifted, dissolved and karstified above the pre-glacial groundwater level, and the paleo-geography of the diagenetically altered Chalk deposits may have been severely disturbed during block-faulting and differential lateral motion of faultblocks. The recording of the present distribution of Chalk facies by means of a detailed regional and supra-regional stratigraphy alone, may not provide a correct understanding of the Late Cretaceous environment of NW Europe. A reconstruction of the depositional and early-diagenetic history of the Late Cretaceous sequences formed on different fault-blocks seems necessary.

# Conclusion

Since the 19th century the understanding of the depositional and early diagenetic conditions of the Chalk environment has much increased. However, the poor preservation and exposure of the extensive Chalk deposits requires a much more detailed reconstruction of the depositional conditions from facies successions on a small scale. The interpretation of the bedding in Chalk and the definition of a model that relates bedding to sea-level fluctuation, subsurface motion and periodic climatic variation during the Cretaceous, will contribute to a better understanding of the Chalk.

# 3 Early Diagenesis of Chalk

### Abstract

Smectitic clay, glauconite, pyrite, carbonate cement and silica occur concentrated in concentric zones around several m high and mm wide U-shaped burrows of *Bathichnus paramoudrae*, common in the Late Cretaceous Chalk of NW Europe. The minerals are authigenic and formed during bacterial metabolism and the decomposition of organic matter in 5 redox zones from aerobic near the burrow wall to anoxic in the surrounding deep sediment. The identification of the succession of redox zones and their specific mineral authigenesis contributes to a better definition of the early diagenetic conditions during deposition of bedded Chalk sequences characterised by a rhythmic vertical variation of the concentrations of the mentioned authigenic minerals.

## Introduction

Smectitic (illite-montmorillonite) clay, glauconite, pyrite, carbonate cement (lithification) and silica (flint) are common minerals in the Late Cretaceous coccolithic mudstones (Chalk) and fine-grained calcarenites (Tuffaceous Chalk) of NW Europe (Håkansson et al., 1974). The minerals are concentrated in layers that parallel erosion surfaces, and that define the rhythmic bedding in the Chalk.

The minerals were formed early, because they occur concentrated in concentric zones around the mm-wide and metres long tubes of deep burrowers, for the first time correctly defined by W.M. Felder (1971) and later thoroughly investigated and named *Bathichnus paramoudrae* by Bromley et al. (1975). For the reconstruction of the depositional/early diagenetic conditions during the sedimentation of the Chalk it is important to understand the genesis of these minerals. Complex processes influenced their distribution below the former sediment surface, such as bioturbative mixing, reworking, transport and sorting and the chemistry of the pore

Figure 3.1 - About 35 m of Chalk of the Gulpen Formation (Lixhe 2-, Lixhe 3- and Lanaye Members) exposed in the southernmost part of quarry ENCI (Maastricht, The Netherlands). Note the thickening upwards of planar, parallel-bedded flint nodule layers, Boirs and Nivelle horizons bounding the Members (large arrows) and abundant *Bathichnus paramoudrae* (small arrows) in the Chalk at the top of the Lixhe 3 Member, below the Nivelle Horizon.

fluid during early diagenetic dissolution, diffusion and precipitation. However, around the deep burrow tubes, mineral authigenesis was not influenced by bioturbative mixing and reworking. Therefore, the authigenesis of minerals in Chalk will be discussed by paying attention to a particularly well preserved specimen of *Bathichnus paramoudrae* in the Chalk of the Lixhe 3 Member (Gulpen Formation, Late Maastrichtian, Felder, 1975a,b), exposed in quarry ENCI (Maastricht).

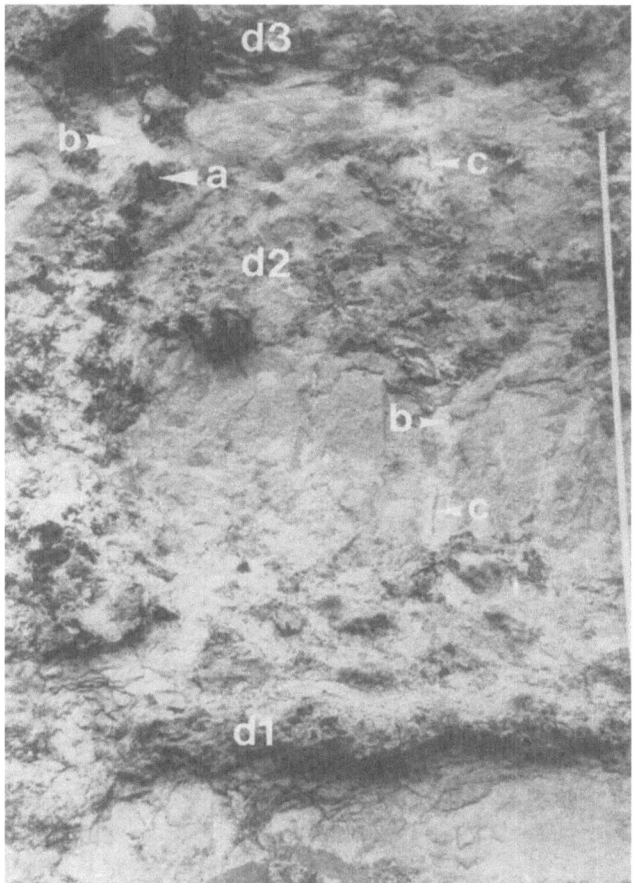

Figure 3.2 - An about 2 m high U-shaped burrow of *Bathichnus paramoudrae* in the Chalk at the top of the Lixhe 3 Member. Well developed flint annulus (a) surrounds the left branch of a U-shaped cylinder of lithified Chalk (b). The glauconitic burrow tube (c) is present in the left branch, turns through the lower flint nodule layer (d1) and proceeds its course upwards in a less well developed right branch, while passing a middle flint nodule layer (d2). The top of the burrow has been eroded or destroyed by bioturbative mixing. Both branches end in the upper flint nodule layer (d3).

## 3.1 The Chalk of the Lixhe 3 Member

The Chalk of the Lixhe 3 Member is a pure coccolithic wackestone with silt-sized bioclasts. The lithofacies is characterized by horizontal silica concretion layers in a

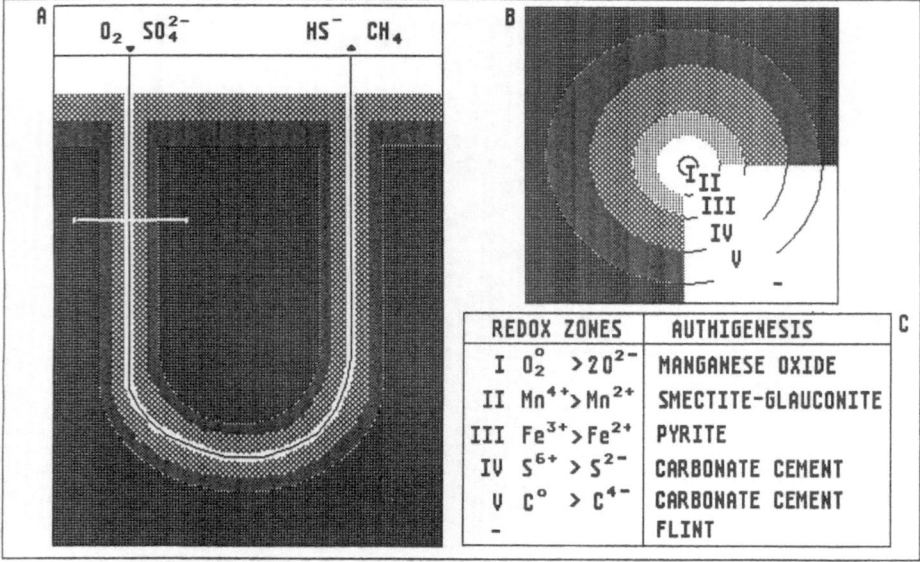

Figure 3.3 - U-shaped burrow of *Bathichnus paramoudrae* in situ with the various redox zones (see text). Cross sections (a,b) with redox zones and associated authigenic mineral zones (c).

close parallel succession (Fig. 3.1). Macrofossils (burrowing echinoids) are rare in the Chalk, that has been homogeneously bioturbated by soft sediment burrowers.

Traces of deep burrows (i.e. *Planolites, Zoophycos, Chondrites* and *Thalassinoides*; Ekdale & Bromley, 1991) are common and moreover silicified. The Chalk of the Lixhe 3 Member, as exposed in quarry ENCI, occurs below the present-day groundwater table. Contrary to younger Chalk, which was oxidized and has a cream-white colour, the Chalk of the Lixhe 3 Member is dark-grey and still contains reduced iron (pyrite, smectite and glauconite). The authigenic minerals within and around the tracefossil *Bathichnus paramoudrae*, common in the Chalk of the Lixhe 3 Member, have been well preserved. Conceivably, when the organism penetrated the Chalk, this was a poorly consolidated watery and muddy mixture of skeletal carbonate, skeletal opal, organic matter and authigenic minerals that formed in redox zones of bacterial metabolism below and parallel to the sediment surface (cf. Hurd, 1973; Håkansson et al., 1974; Schlanger & Douglas, 1974).

## 3.2 *Bathichnus paramoudrae*

*Bathichnus paramoudrae* in the Lixhe 3 Member shows many features that have been considered diagnostic for this trace fossil (Bromley et al, 1975) (Fig. 3.2). The flint nodules that form a ring surrounding the burrow were reported already in the seventeenth century (Worm, 1655; cit. Bromley et al, 1975), and are common in the Cenomanian to Danian Chalk of Northwest Europe. The course of the flint annulus may be straight, winding (sub)vertical or spiral and it may attain a length of more than 6 m (W.M. Felder, 1971). The flint annulus may consist of a few large ring-, pear- or bomb-shaped nodules in a vertical succession, or rarely it consists of one long, single tube. In the studied example, the annulus of flint is 2 m high and it consists of many small nodules.

The flint annulus surrounds the left branch of a U-shaped cylinder of pink-coloured lithified Chalk. A slender green-coloured tube of 1-2 cm in diameter is present in the centre of the lithified Chalk cylinder. The green wall of the tube contains glauconite. The green tube is surrounded by an up to 1 dm wide zone of lithified Chalk, stained by finely dispersed iron-sulphide. Close to the tube the concentration of iron-sulphide is greatest, and it decreases gradually away from the tube.

## 3.3  Earlier interpretations of *Bathichnus paramoudrae*

Without any doubt the above suite of minerals is located around the several mm-wide and several m-long (sub)vertical tube of an exceptionally deep burrower (Bromley et al., 1975). Other mm-wide and m-long burrows were described from the late Campanian Zeven Wegen Member of the lower part of the Gulpen Formation (Lixhe, Belgium) (Bromley, 1968), from the Late Cretaceous Chalk of Mors, Denmark (Nygaard, 1983), and from the Late Cretaceous flysh of Italy (Scholle, 1971).

The nature of the burrower that is held responsible for the trace fossil *Bathichnus paramoudrae*, and the reason why it penetrated the sediment to such great depths has not been explained convincingly in the literature. Bromley et al. (1975) exclude a sediment consuming organism, and suggest that *Bathichnus paramoudrae* was a sheltering tube for a stationary filter feeder or predator. They mention several modern organisms that have the slenderness and the length required to produce *Bathichnus paramoudrae*. Without exception these are vermiform animals that belong to the groups of polychaete annelids, pogonophores and nemertineans. For instance, in the present-day North Sea, a nemertinean, *Heterolineus longissimus*, occurs, which has a width of 2-9 mm and a length of up to 30 m (Hyman, 1951; Stresemann, 1970).

With regard to the early diagenetic processes that led to the concentration of authigenic minerals around the burrow tube, Bromley et al. (1975) advocate the decomposition of the organic matter of the dead burrower and of the organic matter that was introduced during its life as the main cause. Their model follows an earlier

| Redox-zones | Reaction | $\Delta G^{\circ}$ $(KJmol^{-1}CH_2O^*)$ |
|---|---|---|
| Oxic    I | $CH_2O + O_2 \gg CO_2 + H_2O$ | -475 |
| | $5\,CH_2O + 4\,NO_3^- \gg 2\,N_2 + 4\,HCO_3^- + CO_2 + 3\,H_2O$ | -448 |
| Sub-oxic    II | $CH_2O + 3\,CO_2 + H_2O + 2\,MnO_2^{**} \gg 2\,Mn^{2+} + 4\,HCO_3^-$ | -349 |
| III | $CH_2O + 7\,CO_2 + 4\,Fe(OH)_3 \gg 4\,Fe^{2+} + 8\,HCO_3^- + 3\,H_2O$ | -114 |
| An-oxic    IV | $2\,CH_2O + SO_4^{2-} \gg H_2S + 2\,HCO_3^-$ | -77 |
| V | $2\,CH_2O \gg CH_4 + CO_2$ | (-58) |

\* Sucrose   \*\* Birnessite   Data from Latimer (1952), Berner (1971)

Figure 3.4 - Bacterial reactions with decreasing free energy yield in the succession of redox zones, i.e., from aerobic to anoxic conditions.

interpretation presented by Håkansson et al. (1974), who stated that flint formation in and around *Thalassinoides* burrows started after silica had been dissolved and was adsorbed by the organic matter that coated the burrow wall, as was proposed earlier by Siever (1962). However, contrary to the flint related to *Thalassinoides* burrows, flint around *Bathichnus paramoudrae* typically occurs at a distance of several cm to dm from the presumably organic matter-rich central burrow tube.

Other models have been proposed to explain the occurrence of *Bathichnus paramoudrae* in the Chalk of NW Europe. Jordan (1981) considered these structures to be the result of submarine gas or mud volcanoes. Gas or fluids, upwelling from below, would have created an escape track and a chemical gradient that caused the diagenetic corona. This can, however, not have been the case, given the U-shape (Fig. 3.2, 3.3) and laminated internal fill of the burrow tube (Bromley et al., 1975), which clearly demonstrates the bioturbative origin.

Nygaard (1983) described burrows from redeposited Late Cretaceous Chalk above the Mors Salt Dome (Denmark), which he considered to be *Bathichnus* sp. and escape traces. However, *Bathichnus paramoudrae* in the Lixhe 3 Chalk and other Chalk deposits crosses several beds, deposited at rates of presumably only centimetres to decimeters per 1000 years. Therefore these tubes cannot be escape structures.

Clayton (1986) focused attention to the chemical processes that led to the mineralisations around the *Bathichnus paramoudrae* burrow. He adopted part of the

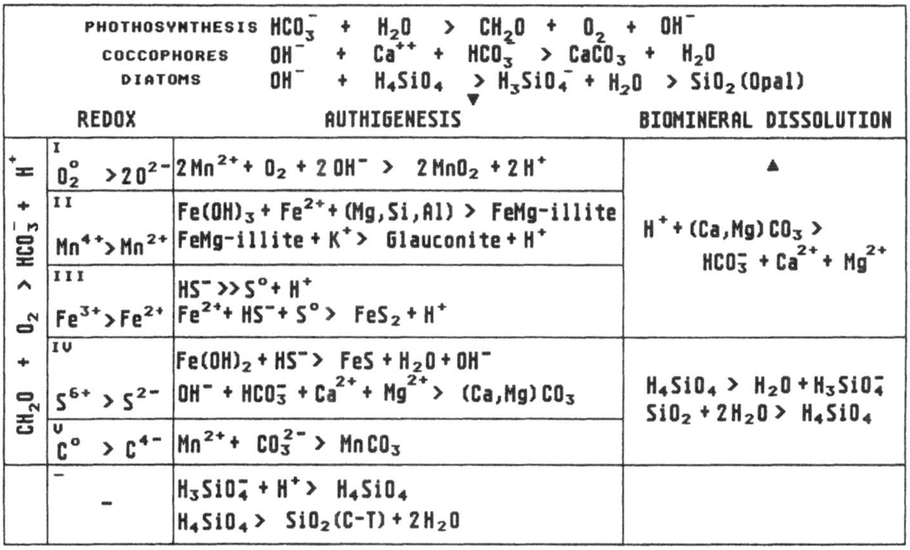

| | | PHOTHOSYNTHESIS $HCO_3^- + H_2O > CH_2O + O_2 + OH^-$ | |
| --- | --- | --- | --- |
| | | COCCOPHORES $OH^- + Ca^{++} + HCO_3^- > CaCO_3 + H_2O$ | |
| | | DIATOMS $OH^- + H_4SiO_4 > H_3SiO_4^- + H_2O > SiO_2 (Opal)$ | |
| | REDOX | AUTHIGENESIS | BIOMINERAL DISSOLUTION |
| $H^+ > HCO_3^- > O_2 > CH_2O$ | I $O_2^0 > 2O^{2-}$ | $2Mn^{2+} + O_2 + 2OH^- > 2MnO_2 + 2H^+$ | |
| | II $Mn^{4+} > Mn^{2+}$ | $Fe(OH)_3 + Fe^{2+} + (Mg,Si,Al) > FeMg-illite$  $FeMg-illite + K^+ > Glauconite + H^+$ | $H^+ + (Ca,Mg)CO_3 >$  $HCO_3^- + Ca^{2+} + Mg^{2+}$ |
| | III $Fe^{3+} > Fe^{2+}$ | $HS^- >> S^0 + H^+$  $Fe^{2+} + HS^- + S^0 > FeS_2 + H^+$ | |
| | IV $S^{6+} > S^{2-}$ | $Fe(OH)_2 + HS^- > FeS + H_2O + OH^-$  $OH^- + HCO_3^- + Ca^{2+} + Mg^{2+} > (Ca,Mg)CO_3$ | $H_4SiO_4 > H_2O + H_3SiO_4^-$  $SiO_2 + 2H_2O > H_4SiO_4$ |
| | V $C^0 > C^{4-}$ | $Mn^{2+} + CO_3^{2-} > MnCO_3$ | |
| | − | $H_3SiO_4^- + H^+ > H_4SiO_4$  $H_4SiO_4 > SiO_2 (C-T) + 2H_2O$ | |

Figure 3.5 - Proposed bacterial metabolism in sediment that was produced during photosynthesis (skeletal carbonate, opal and organic matter). Redox reactions, mineral authigenesis and biomineral dissolution in redox zones in Chalk, based on the observations on *Bathichnus paramoudrae*.

concept of Jordan (1981), assuming that hydrogen-sulphide-rich pore fluids would have risen along the burrow tube and upon oxidation caused the silicification around the burrow. However to allow for his proposed mechanism, a sulphate-rich zone of up to 6 m thickness, i.e., at least as deep as the hight of *Bathichnus paramoudrae*, should have been present. In that case, the Chalk should have been extremely permeable and virtually free of decomposable organic matter. It is, therefore, unlikely that the tube of *Bathichnus paramoudrae* was substantially more permeable than the surrounding sediment. Furthermore, *Bathichnus paramoudrae* crosses several flint layers, presumably formed at the basis of the sulphate-rich zone below and parallel to the sediment surface. Thus, *Bathichnus paramoudrae*, contradicting earlier suggestions (Clayton, 1986), penetrated hydrogen-sulphide-rich sediments, well below the zone of sulphate reduction.

## 3.4 Re-interpretation of *Bathichnus paramoudrae*

The above contradictions require a redefinition of the processes that produced the mineralisations in and around the trace fossil *Bathichnus paramoudrae* (Fig. 3.3). It is suggested that the animal producing the trace fossil *Bathichnus paramoudrae* penetrated sulphate-free sediment well below the zone of sulphate reduction. I suggest that the burrower nourished itself with bacteria, which it cultivated in its burrow by flushing fresh sea water through the U-shaped burrow. The oxygen and sulphate that were thus supplied, reactivated bacterial metabolism. The bacteria decomposed organic matter present in and around the tube and/or used dissolved organic matter that was introduced by the burrower. The bacterial metabolism and decomposition of organic matter altered the chemical environment in the pore fluid around the burrow and caused redox zones that were characterised by dissolution of detrital minerals, by diffusion, and by the precipitation of authigenic minerals.

## 3.5 Bacterial metabolism and authigenesis

The smectite, glauconite, pyrite, carbonate cement and silica, that occur in concentric zones around the tube of *Bathichnus paramoudrae*, conceivably are authigenic products that have been precipitated as the result of early diagenetic bacterial metabolism. The exact mechanism, involving dissolution, diffusion, adsorption and precipitation may be rather complex. A discussion of the bacterial metabolism and the involved enzyme kinetics is outside the scope of this study. A sequence of reactions was proposed (Latimer, 1952; Redfield et al., 1963; Froelich et al., 1979; Berner, 1980) that may occur within the pore fluid of the sediment and that represents the decomposition of organic matter by bacterial activity, in which dissolved oxygen and oxygen from reducible oxides is used (Fig. 3.4). The succession of reactions in the range from aerobic to anoxic conditions reflects a decrease of redox potential and metabolic free energy yield.

The authigenic minerals in the Chalk, notably the carbonate cement, indicate that bacterial metabolism has affected the pH and/or $pCO_2$ of the pore fluid. Because this does not follow from the reaction equations depicted in figure 3.4, a different succession of reactions is proposed to explain the authigenic mineral precipitation in Chalk (Fig. 3.5).

In all redox zones, the bacterial metabolism resulted in the oxidation of organic matter and an increase of the hydrogen-ion and $HCO_3^-$ concentration (Schlegel, 1992):

$$<CH_2O> + 1/2O_2 \rightarrow CO_2 + 2(H) \quad \text{<CH2O>=simplified notation for organic matter}$$
$$\text{(H)=reduced hydrogen bounded to enzymes}$$
$$CO_2 + H_2O \rightarrow HCO_3^- + H^+$$

The decrease of the molecular oxygen concentration led eventually to the reduction of oxides and the precipitation of authigenic minerals. In accordance with the observed authigenic mineral concentration in concentric zones around the burrow tube of *Bathichnus paramoudrae*, the following redox zones and their characteristic reactions are proposed, from the oxygenated towards the anoxic and finally inert sediment:

### 3.5.1 Aerobic zone with dissolved oxygen

In the burrow, where oxygenated seawater occurred, the bacterial decomposition of organic matter by dissolved oxygen caused the dissolution of carbonate:

$$<CH_2O> + 1/2O_2 \rightarrow CO_2 + 2(H)$$
$$CO_2 + H_2O \rightarrow HCO_3^- + H^+$$
$$CaCO_3 + H^+ \rightarrow Ca^{2+} + HCO_3^-$$

### 3.5.2 Suboxic zone of manganese reduction

Near the burrow wall, where glauconite is found (Fig. 3.3), the dissolved oxygen concentration decreased and reduction of manganese oxides started. Glauconite is thought to form under slightly reducing conditions, (Takahashi, 1939; Burst, 1958; Köster & Kohler, 1973; Odin & Matter, 1981).

It was produced in the laboratory within several days at surface temperature and pressure (Harder, 1978, 1980). He found that under slightly reducing/alkaline conditions ($Mn^{4+}$ reduction) iron(III)-hydroxides precipitated and adsorbed silica, iron(II), aluminium and magnesium, which led to the genesis of iron-magnesium-smectites:

$$2MnO_2 + 2H_2O \rightarrow 2Mn^{2+} + O_2 + 4OH^-$$
$$Fe(OH)_3 + Fe^{2+} + Si(OH)_4 + Al(OH)_3 + Mg^{2+} \rightarrow smectite$$

Subsequently smectite was transformed into glauconite by the adsorption of potassium (Odin & Matter, 1981). Around the tube of *Bathichnus paramoudrae*, carbonate was dissolved and replaced by glauconite, suggesting a possible release of hydrogen-ions during potassium uptake:

$$smectite + K^+ \rightarrow glauconite + H^+$$
$$CaCO_3 + H^+ \rightarrow Ca^{2+} + HCO_3^-$$

$$smectite/glauconite = (Si_{(4-x)}Al_x)_4(Fe^{3+}, Al, Fe^{2+}, Mg)_2O_{10}(OH)_2K_{(x+y)}$$
$$0.2<x<0.6 \quad 0.4<y<0.6 \text{ (Odin \& Matter, 1981)}$$

The redox conditions were characterised by manganese-oxide reduction, but not yet by iron-oxide reduction. The Fe(II), that was adsorbed by Fe(III)hydroxide, had to diffuse from the zone of iron reduction (zone III), towards the zone (II) of manganese reduction and clay-mineral genesis. The soluble Mn(II) diffused towards the aerobic zone (I) where it was oxidized and precipitated as Mn(IV) oxide again and towards the most reducing outer zone (V), where it precipitated as manganese-carbonate cement (Suess, 1979; Berner, 1980; Clayton, 1986), giving the pink colour to the lithified chalk cylinder.

### 3.5.3 Suboxic zone of iron reduction

Outside the zone of $MnO_2$ reduction, Fe(III) was reduced to Fe(II).

$$4Fe(OH)_3 \rightarrow 4Fe(OH)_2 + O_2 + 2H_2O$$

The thus formed Fe(II) reacted with hydrogen-sulphide (Berner, 1969, 1970, 1971) that diffused from the anoxic zone (IV) of sulphate reduction and precipitated as FeS (Pyzic & Sommer, 1981).

$$Fe(OH)_2 + H_2S \rightarrow FeS + 2H_2O$$

The genesis of pyrite nodules around the burrow tube (Bromley et al., 1975) required the presence of neutral sulphur as oxidant (Roy & Trudinger, 1970) or of hydrogen-ions as oxidant (Drobner et al., 1990):

$$H_2S + 2Fe(OH)_3 \rightarrow S° + 2Fe(OH)_2 + 2H_2O$$
$$FeS + S° \rightarrow FeS_2$$
or
$$FeS + H_2S \rightarrow FeS_2 + H_2$$

The dissolution of carbonate during the genesis of pyrite nodules may have been caused by sulphur-oxidizing thiobacilli, producing sulphate and hydrogen-ions (Schlegel, 1992).

### 3.5.4 Anoxic zone of sulphate reduction

Outside the suboxic zone (III), sulphate reducing bacteria caused the precipitation of carbonate (Hudson, 1977; Schlegel, 1992):

$$Ca^{2+} + SO_4^{2-} + 8(H) + CO_2 \rightarrow H_2S + 3H_2O + CaCO_3$$

### 3.5.5 Anoxic zone of carbondioxide reduction

Outside the zone of sulphate reduction, an outermost redox zone (V) occurred, where sulphate was depleted and where fermentation (clostridia) occurred together with carbondioxide reduction and methane production (methanogens) (Berner, 1980; Reeburgh, 1980; Woese, 1981; Gautier & Claypool, 1984; Schlegel, 1992).

$$2CO_2 + 8(H) \rightarrow CH_3\text{-}COOH + 2H_2O$$
$$CO_2 + 8(H) \rightarrow CH_4 + 2H_2O$$

The decrease of the carbondioxide concentration led to an increase of the hydroxyl-ion concentration.

$$HCO_3^- \rightarrow CO_2 + OH^-$$

The increase of the hydroxyl-ion concentration was immediately buffered and this had two effects on the sediment. On the one hand it resulted in the precipitation of otherwise rather soluble manganese carbonates.

$$HCO_3^- + OH^- + Mn^{2+} \rightarrow MnCO_3 + H_2O$$

On the other hand, in the presence of skeletal opal and monomeric silicic acid, the increase of pH was buffered by the dissociation of the monomeric silicic acid.

$$H_4SiO_4 + OH^- \rightarrow H_3SiO_4^- + H_2O$$

Alkaline conditions in the anoxic redox zones were also suggested on the basis of studies of recent anoxic sediments (Suess, 1979), where manganese-carbonate coatings formed around bacterial colonies, that were embedded in a spongy matrix of silica.

The negative silica complexes, formed during carbondioxide reduction, did not precipitate on the negatively charged skeletal opal surface (Parks, 1967), but accumulated in the pore fluid, while skeletal opal further dissolved in order to maintain the monomeric silicic acid saturation concentration (Williams et al., 1985; Williams & Crerar, 1985).

$$SiO_2 \text{ (Opal)} + 2H_2O \rightarrow H_4SiO_4$$

Negatively charged complexes may have polymerized (Baes & Mesmer, 1976; Iler, 1973, 1979) and precipitated as silica polymorphs, less soluble than skeletal opal, or they may have diffused towards the sediment outside the carbondioxide reduction zone, where they associated with hydrogen-ions again, and thus super-saturated the pore fluid with respect to the monomeric silicic-acid saturation concentration and then precipitated as a less soluble silica polymorph. During late diagenesis, all the skeletal opal in the surrounding sediment dissolved and silica was concentrated around the less soluble polymorphs, causing the growth of the flint nodules (Knauth, 1979; Voigt, 1979a; Zijlstra, 1987, 1989).

# Discussion

The U-shape of *Bathichnus paramoudrae*, the difference between the degree of authigenesis around both arms and the well developed concentric zones of authigenic mineral concentration at some distance from the burrow tube, suggest that the burrower flushed seawater through its burrow in order to generate a redox-potential and thus actively stimulated the growth of bacteria that it subsequently harvested.

Several reactions have been proposed above to explain the genesis of authigenic minerals around the burrow tube of *Bathichnus paramoudrae* in 5 redox zones from aerobic near the burrow wall to anoxic conditions in the sediment around. These reactions should be considered as a short and simplified notation for the complex early diagenetic processes that have altered the Chalk sediment. Understanding of the properties of complex ecosystems of mutually dependent bacterial species (van Gemerden, 1993), as occurred around the burrow of *Bathichnus paramoudrae*, is far from complete. It is to be expected that bacteria prefer to perform their metabolism with a minimal deviation from the thermodynamic equilibrium. Therefore, bacteria may remove or add considerable amounts of reactants and products from the pore fluid only if a deviation from the thermodynamic equilibrium is buffered by the dissolution of detrital minerals and/or the precipitation of authigenic minerals.

The succession of authigenic mineral zones around the burrow tube hardly allows the reconstruction of the dynamics of the early diagenetic processes. However, in order to explain the presence of Fe(III) as a source for Fe(II) in the sub-oxic zone of iron-reduction, it is suggested that when the organism penetrated, first oxygen and sulphate diffused from the oxygenated burrow tube far into the anoxic sediment around. Consequently iron-sulphides were oxidized to Fe(III)-hydroxides and sulphates. When the bacterial activity increased, oxygen and sulphate were exhausted progressively closer to the burrow wall and while the redox zones contracted around the burrow tube, the bulk of the authigenic minerals was precipitated.

The concentrations of authigenic minerals in and around the burrow tube of *Bathichnus paramoudrae* thus stress the importance of the role of bacterial metabolism during early diagenesis and they allow the recognition of different redox zones in their proper succession and with their specific authigenic minerals.

# Conclusion

Conceivably, the several metres deep and mm-wide burrows of *Bathichnus paramoudrae* were constructed in order to generate a redox potential gradient between deep anoxic sediment and oxygenated sea water, introduced and flushed through the U-shaped burrow. The consequent introduction of oxygen and sulphate stimulated bacterial metabolism and the decomposition of organic matter, thus providing food for the burrower.

Bacterial metabolism caused the reduction of oxides and several redox zones formed around the burrow tube. Changes of the thermodynamic equilibrium in the pore fluid were buffered by the dissolution of detrital biominerals (skeletal carbonate and skeletal opal) and the precipitation of authigenic minerals.

The succession of authigenic mineral zones around the burrow reflects the change of redox conditions from oxygenated close to the burrow towards anoxic in the sediment around. Resuming, the authigenic minerals in Chalk were formed in the following redox zones:

**I** Aerobic zone with dissolved oxygen - Carbonate dissolution, manganese-oxide precipitation.

**II** Suboxic zone of manganese-oxide reduction - Carbonate dissolution, smectite precipitation and genesis of glauconite when smectite adsorbed potassium.

**III** Suboxic zone of iron-hydroxide reduction - Carbonate dissolution and iron-sulphide precipitation.

**IV** Anoxic zone of sulphate reduction - Carbonate precipitation

**V** Anoxic zone of carbondioxide reduction - Manganese-carbonate precipitation and/or silica dissolution and silica polymerization.

**VI** Inert zone - Silica precipitation

# 4 Origin and Growth of Flint Nodules

### Abstract

Cryptocrystalline quartz concretions (flint) of different density, size, morphology and abundance, occur isolated and concentrated in layers in the Late Cretaceous (Tuffaceous) Chalk of NW Europe. Concretions started to grow during early diagenesis at sites of elevated authigenic silica-polymorph concentrations, a result of bacterial metabolism in an anoxic mixture of skeletal carbonate, skeletal opal, organic matter and pore fluid. The late diagenetic growth of flint nodules has been investigated with the help of a numerical model that simulates the dissolution, diffusion and precipitation of different silica polymorphs in a medium of changing porosity/permeability. It is suggested that the occurrence of various flint types reflects the distribution of the early diagenetic, authigenic silica polymorphs, rather than the distribution of detrital skeletal opal. The model results imply that flint nodules can be used to reconstruct the depositional and early diagenetic conditions during the genesis of sequences of Chalk with flint.

## Introduction

Cryptocrystalline quartz concretions (flint) (Buurman & van der Plas, 1971) are common in the Campanian-Maastrichtian porous coccolithic mudstones (Chalk) and bioclastic siltstones (Tuffaceous Chalk) of Maastricht. At Rijkholt near Maastricht flint was already excavated during the Neolithic in order to produce tools and weapons, and flint nodule layers have been used for lithostratigraphical correlation since the 18th century (Binkhorst van den Binkhorst, 1859, Felder, 1975 a,b).

The karstified (Albers & Felder, 1979) Maastrichtian carbonates of Maastricht form a 120 m thick coarsening upwards sequence. The size, form and distribution of flint nodules correlates with the lithostratigraphy of the (Tuffaceous) Chalk

52

Figure 4.1 - Dense nodules
with a thin porous margin
in 3 m of pure coccolithic
bioclastic siltstones of the
Lanaye Member (arrow points
to erosion surface) (quarry
ENCI, Gulpen Formation,
Late Maastrichtian).

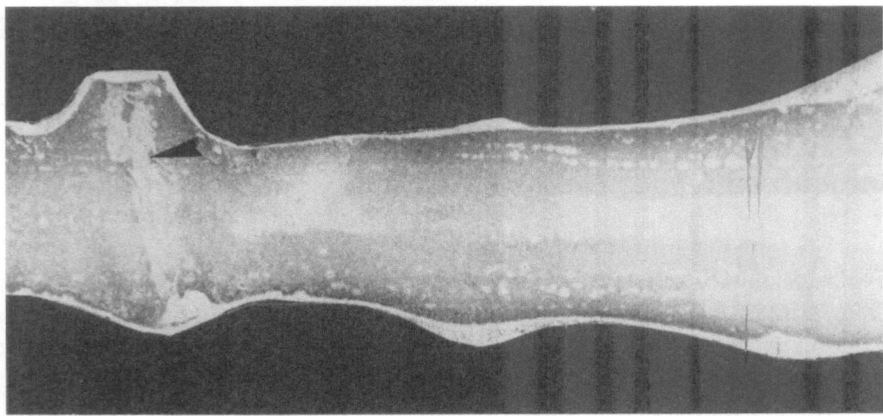

Figure 4.2 - Platy silica concretion (width 20 cm) with porous core and dense margins (arrow points
to burrow), in pure laminated bioclastic sand of the Emael Member (quarry ENCI, Maastricht
Formation, Late Maastrichtian).

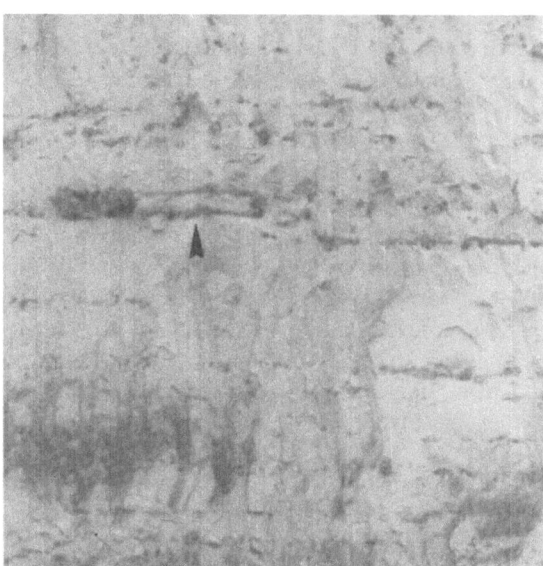

Figure 4.3 - Large silica nodule (width 200 cm, arrow) with non-silicified core and dense margins, in 7 m of pure, partly laminated, bioclastic sand of the Emael Member (quarry ENCI, Maastricht Formation, Late Maastrichtian).

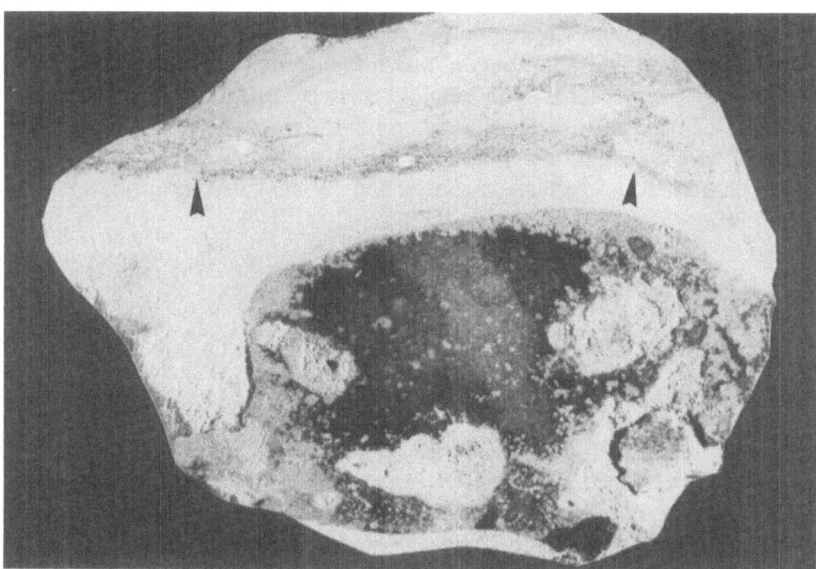

Figure 4.4 - Silica concretion (width 15 cm) with dense core and thick porous margins, in slightly smectitic/glauconitic bioclastic silt of the top of the Valkenburg Member.
Arrows point to St Pieter Horizon that is covered by coarse-grained glauconitic, bioclastic sand.
Note possible early diagenetic lithification and de-silicification as indicated by the erosion surface, parallel to the porous nodule margin (quarry ENCI, Maastricht Formation, Late Maastrichtian).

(Felder, 1975). Clayey (smectitic) coccolithic mudstones (Vylen Member, Gulpen Formation) occur at the basis of the sequence and are characterised by isolated, low density flint nodules with a morphology that still reflects their origin and genesis in and around burrows (Bromley & Ekdale, 1986). The smectitic Chalk changes gradually upwards into pure coccolithic mudstones and bioclastic-coccolithic siltstones (Lixhe 1,2,3 and Lanaye Members, Gulpen Formation). Flint nodules are abundant, dense, irregular and concentrated in layers, while moreover the initial burrow morphology has been poorly preserved. The Chalk with Flint changes upwards into bioclastic sandstones (Tuffaceous Chalk, Maastricht Formation). Flint nodules are common in the lower part of the Tuffaceous Chalk and moreover they are platy in wavy laminated bioclastic sandstones (Felder, 1975a,b).

The high concentration of the flint, in particular in the rather pure coccolithic-rich carbonates of the Gulpen Formation, suggests that the silica of flint was derived from fine-grained detrital skeletal opal (Soudry et al., 1981). The skeletal silica that dissolved, concentrated at centres of precipitation in the former sea bottom (Ehrenberg, 1812; Calvert, 1974; Voigt, 1979a). It has been shown that the centres of precipitation were located in or just outside the most anoxic redox zones of sulphate and/or carbondioxide reduction (Chapter 3, Zijlstra, 1987, 1989). These zones were wrapped around burrows in deep inert sediment and occurred at some depth below and parallel to the former sea bottom.

Here, attention will be paid to the late diagenetic growth of the silica concretions around the early diagenetic precipitation centres, in order to determine whether flint nodule morphology and distribution provide information about depositional and early diagenetic conditions. The relation between flint nodules and the initial distribution of detrital skeletal opal and/or authigenic silica precipitates is investigated with the help of a numerical model that simulates dissolution, diffusion and precipitation in a medium, characterised by porosity/permeability changes in space and time.

## 4.1 The morphology of flint nodules

Flint nodules of the size of a pigeon egg to a millstone are found in the carbonates of Maastricht. The form of the nodules is tubular or complex regular, when they formed in and around burrows (Bromley & Ekdale, 1986). Where the degree of silicification is high, the nodules have amalgamated into irregular lumps of silica (Felder, 1975a). In laminated sediments without traces of bioturbation, the flint nodules are platy. Furthermore, one may find pebbles of flint overgrown with Maastrichtian fossils (P.J. Felder, 1971; Voigt, 1979a) and silicified molluscs (Schins & Buurman, 1979), bryozoans, driftwood and sea grass (Voigt & Domke, 1955).

The colour of flint nodules varies between bluish-black, bluish-grey and white.

The darker the flint, the less light it reflects. Dark flints are saturated with silica and represent dense silicifications. Lighter coloured flints are porous and reflect a substantial portion of the light at silica/carbonate/porespace boundaries. The density of the flint nodules is related to the lithology of the host rock (Felder, 1975), and may vary within a single nodule.

One may distinguish three nodule types in the Tuffaceous Chalk sequence: type 1 - dense nodules with a thin porous rind, that occur in pure Chalk (Fig. 4.1); type 2 - porous nodules with a dense margin in Tuffaceous Chalk (Fig. 4.2, 4.3); and type 3 - porous nodules with one or more dense cores, that are common in smectitic/ glauconitic Tuffaceous Chalk (Fig. 4.4).

## 4.2  The chemistry of silica in calcareous sediment

Skeletal opal with a high surface area (Kamatani & Riley, 1979) dissolves relatively rapidly in neutral water and forms monomeric silicic acid, when the $SiO_2$ molecule combines with two water molecules (Okamoto et al, 1957).

$$SiO_2 + 2H_2O \rightarrow H_4SiO_4$$

In the presence of a proton acceptor (base), the monomeric silicic acid dissociates and a negatively charged complex is formed (Baes & Mesmer, 1976). During early diagenesis, bacteria reduce sulphate and carbondioxide at some depth below the seabottom and they increase the hydroxyl-ion concentration. The increase of the pH is buffered by monomeric silicic-acid dissociation (Zijlstra, 1987, 1989). For instance in the outermost redox zone of carbondioxide reduction and methane genesis with enzymatically bounded hydrogen oxidant ((H) Schlegel, 1992; Chapter 3):

$$CO_2 + 8(H) \rightarrow CH_4 + 2H_2O$$
$$HCO_3^- \rightarrow CO_2 + OH^-$$
$$H_4SiO_4 + OH^- \rightarrow H_3SiO_4^- + H_2O$$

The negatively charged complex does not precipitate on the negatively charged solid silica surface (Parks, 1967). Therefore, during continued bacterial metabolism, the negatively charged complexes accumulate in the pore fluid, while solid silica further dissolves in order to maintain the monomeric silicic-acid saturation concentration.

The negatively charged complexes may polymerize (Iler, 1973, 1979) and precipitate as new silica solid with a lower solubility than the original skeletal opal (Williams et al., 1985; Williams & Crerar, 1985). If the fluid becomes less basic, then the negatively charged complexes may associate with hydrogen ions again, the fluid becomes supersaturated with respect to the saturation concentration of the monomeric silicic acid, and also in this case authigenic silica precipitates.

The solubility of silica is a function of the hydrogen-ion concentration (pH) and is furthermore influenced by the reaction of silica with metals, complexation (Fournier & Marshall, 1983), adsorption (Siever & Woodford, 1973) and neoformation (Harder, 1980; Drever, 1982). In the pore fluid of pure carbonates, the hydrogen-ion concentration is considered the most important factor for the silica solubility. The availability of hydrogen ions for reaction with silica complexes and/or (hydrogen) carbonates depends on the carbondioxide solubility. During late diagenesis, organic matter oxidation and temperature/pressure variations caused variations of carbondioxide pressure and silica/carbonate solubility. An increase of the $pCO_2$ leads to carbonate dissolution and silica precipitation and a decrease to carbonate precipitation and silica dissolution.

$$CaCO_3 + CO_2 + H_2O \longleftrightarrow Ca^{2+} + 2HCO_3^-$$
$$H_3SiO_4^- + CO_2 + H_2O \longleftrightarrow HCO_3^- + H_4SiO_4$$
$$H_4SiO_4 \longleftrightarrow SiO_2 + 2H_2O$$

The variation of the carbondioxide pressure and acidity of the pore fluid during late diagenetic meteoric/groundwater percolation influenced the reaction rates and the rate of flint nodule growth (Albers & Felder, 1979; Knauth, 1979). However, the nodule morphology was mainly a function of the local spatial and temporal variations of the solid silica surface solubility (Krauskopf, 1959), of the variation of the saturation concentrations and of the dissolved silica diffusion rates (Applin, 1987).

The silica solubility is a function of the order (crystallinity) of the arrangement of silicium and oxygen atoms at the solid surface. One may distinguish several silica polymorphs with an increasing crystallinity and decreasing solubility, expressed by the decrease of the saturation concentrations (i.e., Opal, 60-130 ppm; Cristobalite-Tridymite, 20-30 ppm and Quartz 6-10 ppm at 25° C and pH 7; Iler, 1979).

The degree of crystallinity increases in time (Kästner et al. 1977), and is partly defined during the continuous dissolution and precipitation at the surface of the solid, and partly by a much slower later re-arrangement of atoms in the poorly ordered solid silica (solid-solid conversion) (Zemmels & Cook, 1973; Wise & Weaver, 1974).

The crystallinity of the solid surface depends on the precipitation rate that is proportional to the saturation concentration. The crystallinity is low if the molecules accumulate rapidly. On the contrary, if the monomeric silicic acid concentration and the precipitation rate are low, then the poorly ordered solid surface will dissolve more rapidly than the better ordered surface. Slow precipitation at low concentrations causes the genesis of well ordered polymorphs, that have lower saturation concentrations and that give way to the precipitation of even better ordered polymorphs. Silica precipitation is autocatalytic (Krauskopf, 1959).

After early diagenesis, part of the porous carbonate rock contained soluble skeletal opal with a high saturation concentration and part was characterised by a certain concentration of less soluble early diagenetic silica polymorphs with a lower saturation concentration, formed during bacterial metabolism (Chapter 3; Zijlstra, 1987, 1989). A spatial variation of saturation concentrations, as a result of different skeletal opal/authigenic polymorph ratios, caused a concentration gradient which led to diffusion of dissolved silica. Diffusion decreased the concentration gradient and in the vicinity of skeletal opal the pore fluid thus became undersaturated with respect to the opal saturation concentration and opal dissolved. In the vicinity of the less soluble authigenic polymorphs, the pore fluid became supersaturated relative to the saturation concentration of these polymorphs and precipitation occurred. The autocatalytic precipitation of silica guaranteed the continuous dissolution of the least ordered polymorphs and the diffusion of dissolved silica towards and precipitation at the sites of the best ordered polymorphs, until a dense silica concretion was formed in a porous carbonate hostrock from which ultimately all silica has been dissolved.

While polymorphs dissolved and precipitated, porosity and permeability changed, so did diffusion rates. The relation between dissolution, precipitation and diffusion in a medium of which the pore space and the diffusion rates changed during dissolution and precipitation, is further investigated with a self-organizing numerical model (Wolfram, 1986).

## 4.3  The numerical simulation of flint nodule growth

The simulation of flint nodule growth is performed with a numerical model that changes an initially diffuse distribution of authigenic silica, present in a series of connected unit volumes, into a dense concentration of silica in a few volumes of initially highest concentration. The model must allow the simulation of the genesis of concretions with a dense core and porous margins (type 1) and of concretions with a porous core and dense margins (type 2).

The model records the change of reaction rates as a function of the change of the solid silica volume, surface area and solubility, and of the change of the dissolved silica concentrations in the adjacent unit volumes. Furthermore, it records the change of the dissolved silica diffusion rates as a function of changing concentration differences and of changes of permeability (flux area) between connected volumes.

Initially, one defines a series of N connected unit volumes, cubes with sides 1 and volume 1. Solid silica is present as spheres with their centres at the centres of the cubes. When the diameter of the spheres is larger than the side of a cube and when the spheres occur partly outside the cube, then the part outside the cube is

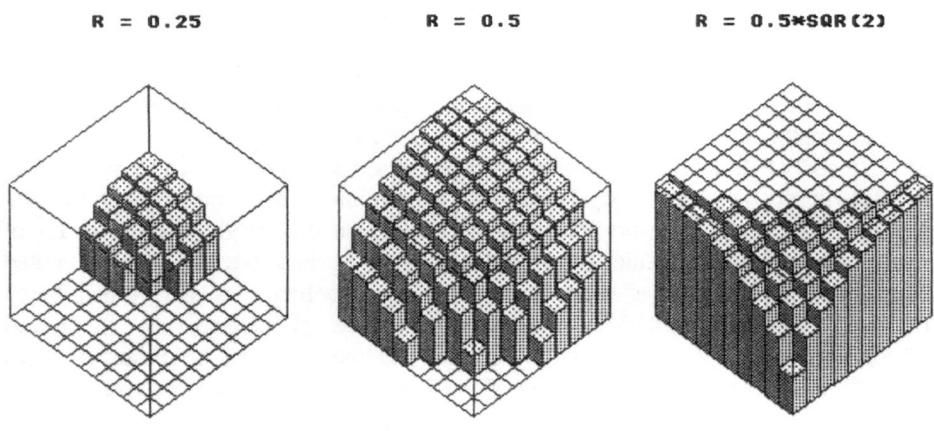

Figure 4.5 - 1/8 th of the simulation cube with solid silica volume, surface area and flux area for increasing radius of the sphere confined in the cube.

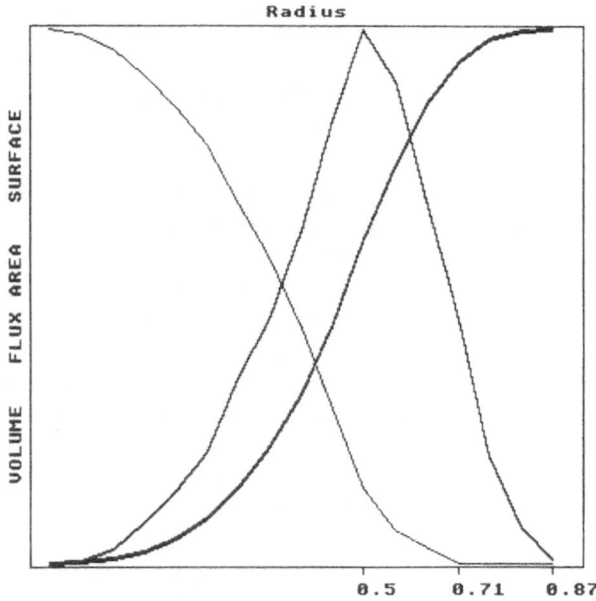

Figure 4.6 - Solid volume (thick line), surface area (intermediate line) and flux area (thin line) change for increasing radius of a sphere confined within a cube.

neglected. It is assumed that if spheres grow outside the cube they connect with spheres of adjacent volumes. The pore fluid equals 1 minus the solid silica volume (porosity). The size of the solid surface area, that controls the reaction rate, is the surface of the sphere confined within the cube. The flux area, that controls the diffusion rate, is 1 minus the largest cross section of the sphere within the cube. For the sake of simplicity, the cube contains only solid silica and no solid carbonate (Fig. 4.5).

For a number of spheres with an increasing radius (R), the solid volume (S), surface area (A) and flux area (F), have been calculated (Fig. 4.6). During precipitation and growth of the spheres (increasing R) the volume increases, the flux area decreases and the surface area increases as long as R<0.5. If R further increases and the sphere tends to grow outside the cube, then the surface area decreases again. Only diffusion from adjacent cubes, but no further diffusion through a cube, can occur when R>0.5*SQR(2) (0.71) and the flux area is zero. The cube is entirely filled with solid silica when R=0.5*SQR(3) (0.87).

In the numerical simulation model for nodule growth, first (t=0) the initial solid volume distribution (procedure 1), the solid surface solubility (procedure 2) and the dissolved saturation concentration (procedure 3) are defined. During the simulation time (T) reaction (procedure 4) alternates with diffusion (procedures 5-I, 5-II).

+ = addition
- = substraction
* = multiplication
/ = division
SQR( ) = square root
sin( ) = goniometric function
A(n) = an array of n sites
m^n = m to the power n
pi = 3.14..

```
FOR t=0 TO t=T (step dt)
   IF t=0
           GO TO PROCEDURE 1 (initial solid volume distribution)
           GO TO PROCEDURE 2 (initial solid surface solubility)
           GO TO PROCEDURE 3 (initial dissolved concentration)
   ELSE
           GO TO PROCEDURE 4 (reaction)
           GO TO PROCEDURE 5-I (fast diffusion)
                   or
           GO TO PROCEDURE 5-II (permeability dependent diffusion)
   ENDIF
NEXT t
```

## 4.3.1 Initial conditions

PROCEDURE 1 - Initially the N connected cubes are filled with solid volume, in such way that from cube 0 to cube N-1. the solid concentration (S(i)) increases and decreases again, simulating the distribution of an early diagenetic polymorph. The defined solid volume S(i) allows the definition of the solid surface areas A(i) and the flux area F(i) in the N cubes (Fig. 4.6).

```
FOR i=0 TO i=N-1
     S(i) = Smax*(0.5 + 0.5*sin(((2*pi)/N)*i-0.5*pi))
NEXT i
```

S(i) = the solid concentration in the ith volume 0<S(i)<Smax
Smax = the maximum initial solid concentration 0<Smax<1
RETURN

PROCEDURE 2 - The solubility of the solid surface area (Sol(i)) in the ith cube, that is the dissolution rate per unit surface as a function of crystallinity, is initially the same in each volume.

```
FOR i=0 TO i=N-1
     Sol(i) = Sol0
NEXT i
```

Sol(i) = the solubility of the solid in the ith volume
Sol0 = the initial solubility 0<Sol0<1
RETURN

PROCEDURE 3 - The concentration (C(i)) of dissolved silica in the pore fluid (1-S(i)) of the ith volume is also the same everywhere and initially it is slightly higher than the solubility Sol(i), in order to provoke precipitation.

```
FOR i=0 TO i=N-1
     C(i) = C0
NEXT i
```

C(i) = the concentration of dissolved silica in the ith volume
C0 = the initial concentration 0<C0<1 C0>Sol0
RETURN

## 4.3.2 Reaction

PROCEDURE 4 - Reaction occurs when the solid surface solubility (Sol(i)) and the dissolved concentration (C(i)), are different. The volume of the solid is changed by precipitation or dissolution. The amount of solid volume change increases with increasing solid surface (A(i)) and with increasing difference between solid solubility and dissolved concentration (Sol(i)-C(i)).

```
FOR i=0 TO i=N-1
     S(i) = S(i)+(A(i)/Amax)*(C(i)-Sol(i))
     C(i) = C(i)-(A(i)/Amax)*(C(i)-Sol(i))
NEXT i
```

A(i) = solid surface area
Amax = maximum solid surface area (4*pi*0.5^2)

After dissolution or precipitation, the concentration of dissolved silica is more in accordance with the solubility of the old solid surface. The new solubility Sol(i) of the exposed (dissolution) or newly formed (precipitation) solid surface is a function of the old solubility and the volume change of solid silica.

```
FOR i=0 TO i=N-1
        Sol(i) = Sol(i)-f*(A(i)/Amax)*Sol(i)
NEXT i
```

f = proportionality factor 0<f<1

The procedure decreases the solubility least if solid has been dissolved (poorly ordered polymorph dissolution), while it decreases the solubility most if solid has been formed and the solid surface area increased (better ordered polymorph precipitation).
RETURN


## 4.3.3  Diffusion

PROCEDURE 5-I - Precipitation and dissolution produce a new solid surface solubility distribution and a dissolved silica concentration gradient. The concentration gradient provokes diffusion. The degree of dissolved matter exchange between adjacent volumes is a function of the brownian motion of the dissolved silica molecules, the concentration difference between adjacent volumes and the smallest of the two flux areas in the adjacent volumes. Very fast diffusion, disregarding flux area, is simulated by the following procedure:

```
FOR i=0 TO i=N-1
        sumC = sumC + C(i)
NEXT i
FOR i=0 TO i=N-1
        C(i) = sumC/N
NEXT i
RETURN
```

PROCEDURE 5-II - Considering flux area during diffusion, the following procedure is used:

```
FOR t=0 TO t=dt
        FOR i=0 TO i=N-1
                IF i=0
                        i-1=N-1
                ENDIF
                IF i=N-1
                        i+1=0
                ENDIF
                IF F(i)>F(i-1)
                        Da=F(i-1)*v*0.25
                ENDIF
                IF F(i)=F(i-1) OR F(i)<F(i-1)
                        Da=F(i)*v*0.25
```

```
                ENDIF
                IF F(i)>F(i+1)
                        Db=F(i+1)*v*0.25
                ENDIF
                IF F(i)=F(i+1) OR F(i)<F(i+1)
                        Db=F(i)*v*0.25
                ENDIF
                H(i)=C(i)-Da*C(i)-Db*C(i)+Da*C(i-1)+Db*C(i+1)
        NEXT i
        FOR i=0 TO i=N-1
                C(i)=H(i)
        NEXT i
  NEXT t

  F(i) = flux area
  Da,Db = diffusion coefficients
  v = velocity of dissolved silica 0<v<1
  H(i) = local array
  t,dt = the diffusion procedure is repeated dt times before the reaction procedure is executed dt>0
RETURN
```

The diffusion rate may be slowed down relative to the reaction rate by decreasing the velocity (v) of diffusing silica. The diffusion rate may be increased relative to the reaction rate if dt is increased and thus the diffusion procedure is repeated more often before the reaction procedure is executed.

In procedure 5-II, the array has been closed by allowing diffusion between the 0 and N-1 cubes. Eventually, during each iteration, the concentration of the 0 and N-1 volumes may be changed slightly, in order to simulate diffusion in and out of the closed simulation space.

The procedures are repeated in order to simulate the development of dense silica concretions. Before each iteration, the new solid surfaces A(i) and flux areas F(i) are defined in accordance with the new solid volumes S(i) and according to the relation between A,F and S for different R as was defined in advance (Fig. 4.6).

### 4.3.4 Simulation results

For a space of 32 volumes (N=32), with an initial solid distribution of Smax=(0.5), initial concentration C0=0.6, initial solubility Sol0=0.5, with factor f=0.1 and dissolved silica velocity v=1, two different simulations of 16 iterations have been performed. In the first simulation a fast diffusion rate, relative to the reaction rate, has been assumed (procedure 5-I). The distribution of the solid appears to change from a smooth porous, to a concentrated dense distribution, with a thin porous rind (Fig. 4.7) (nodule type 1; Fig. 4.1). In the second simulation, the flux area was not neglected and the diffusion rates were slower. The diffusion procedure (5-II) was repeated 100 times (dt=100), before the reaction procedure (4) was executed. The

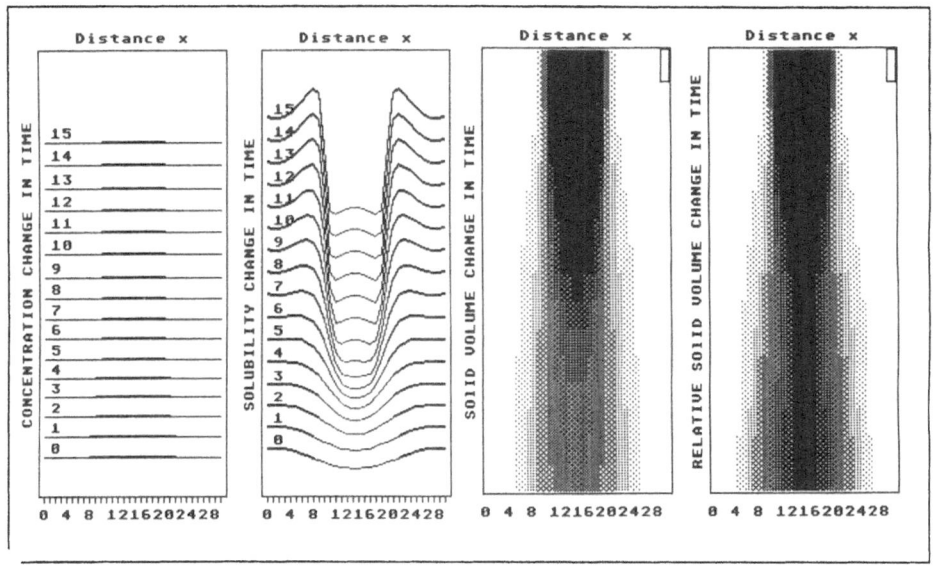

Figure 4.7 - Simulation of silica nodule growth for fast diffusion rates (procedure 5-I). From left to right: the dissolved concentrations, the surface solubilities, the absolute solid concentrations and the solid concentrations relative to adjacent volumes are depicted for a space of 32 cubes, during a simulation period of 16 iterations (bottom to top). Note the initial dispersed concentration of solid and the genesis of a dense core with thin porous margins.

distribution of the solid changed from smooth porous, to a more concentrated dense distribution, which ultimately bifurcated into two dense margins and a central porous core (Fig. 4.8) (nodule type 2; Fig. 4.2, 4.3). Nodule type 3, a porous nodule with one or more dense cores, might be represented by the earliest simulation results, but disappears during the simulation as the distribution evolves to the equivalents of nodule types 1 or 2.

## Discussion

The first simulation (diffusion procedure 5-I) reflects the genesis of dense flint nodules with a thin porous rind that occur in Chalk. The second simulation (diffusion procedure 5-II) reflects the genesis of flint nodules in Tuffaceous Chalk, which may have margins that are more dense than the core.

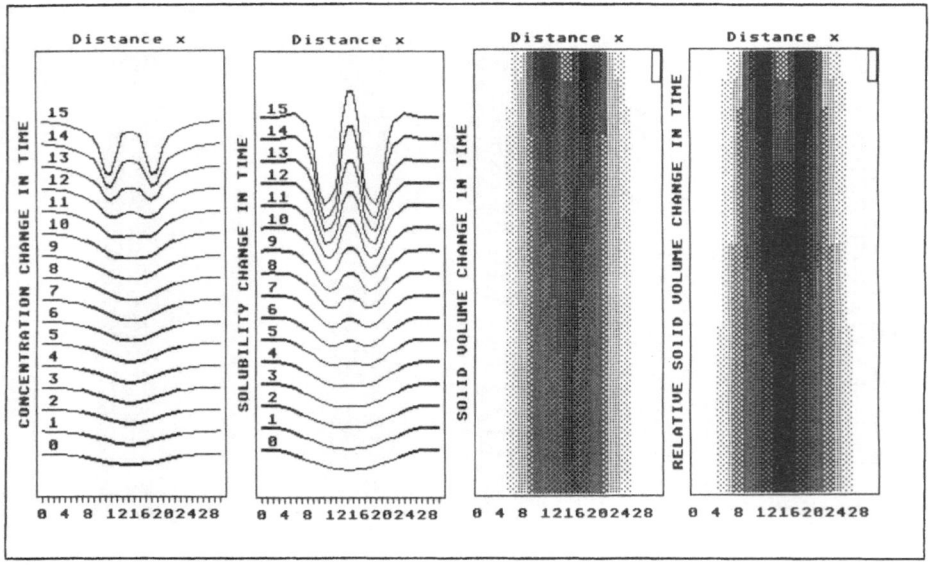

Figure 4.8 - Simulation of silica nodule growth for low diffusion rates (procedure 5-II), depending on flux area. From left to right: see Fig. 4.7. Note the initial dispersed concentration of solid and the genesis of a porous core with dense margins.

The model suggests that differences between both flint nodule types have been caused only by differences in the diffusion rates, relative to the reaction rates. These differences thus would have been mainly a function of the change of the flux area or permeability in the nodule during growth and of the reaction rate defining $pCO_2$ during late diagenetic meteoric groundwater percolation.

The difference between nodule types in coarse-grained Tuffaceous Chalk, fine-grained pure Chalk with Flint and fine-grained smectitic Chalk might thus be explained by the differences in the change of carbonate dissolution rate of the host rock during nodule growth. Although carbonate was neglected in the model, some predictions about the role of the hostrock lithology can be made.

Despite the precipitation of the silica, the porosity, permeability and diffusion rates remained high in the fine-grained pure Chalk, where the carbonate surface exposed to the pore fluid was large and where carbonate dissolved easily. In the coarse-grained Tuffaceous Chalk, the carbonate surface area exposed to the pore fluid was much smaller than in fine-grained Chalk of the same porosity. The dissolution of carbonate proceeded more slowly and the precipitation of silica resulted in a substantial decrease of the porosity, permeability and diffusion rates.

It should also be noted that, additionally, the flow of $CO_2$-rich meteoric groundwater during karst first reached the flint of the Tuffaceous Chalk and because the silica precipitation rate increases with increasing $CO_2$ concentration, a higher $CO_2$ concentration in the Tuffaceous Chalk, as compared to the Chalk below, may have further promoted the genesis of nodules with dense margins and a porous core, instead of dense nodules with a porous margin, like they occur in the Chalk below.

In order to produce the third nodule type, a porous nodule with one or more dense cores, which occurs in smectitic-glauconitic Chalk, one has to arrest the nodule growth before it reaches one of the above shown distributions (Figs. 4.7, 4.8). Nodule growth may stop if the dissolved silica concentration becomes equal to the silica saturation concentration of clay minerals in and around the nodules. In that case, the concentration is the same in the core and margins of the nodule. The concentration gradient is then absent and no further diffusion and nodule growth can occur. Nodule growth may also stop if the surface solubility of the margin and core of the nodule is the same. This condition occurs if the diffusion rates in a compacted clayey carbonate are so low that surface solubility is no longer a function of dissolution/precipitation rates but of solid-solid conversion.

It thus appears, that flint nodule genesis is a function of silica precipitation rates, carbonate dissolution rates and of dissolved silica diffusion rates, depending on both. The initial skeletal-opal distribution is considered of minor importance and was not included in the simulation model. The rate of opal dissolution is much higher than the rate of authigenic silica precipitation. The constant diameter of tubular flint nodules around *Thallassinoides* burrows, and even more convincingly, the rather constant diameter of a 6 m long vertical flint cylinder around the burrow tube of *Bathichnus paramoudrae*, intersecting various flint nodule layers (W.M. Felder, 1971), clearly demonstrate the importance of the early diagenetic silica polymorph concentration and the distribution of its surface area and the unimportance of skeletal opal distribution, as far as it concerns the flint nodule morphology.

The value of the model for flint nodule genesis, presented here, is accentuated by the fact that it explains the genesis of flint nodules with a porous core that is isolated from the carbonate rock surrounding the nodule. These nodules are common in the Tuffaceous Chalk of South Limburg and the preservation of partly silicified carbonate sediment within such nodules, known for its well preserved microfossils (flint meal) (Hart et al., 1986) is thus explained.

## Conclusion

The early to late diagenetic genesis of flint nodules occurs at sites with an elevated concentration of early diagenetic silica polymorphs. According to the presented model, the form and size of the flint nodules reflect the distribution and concentration of the early diagenetic silica polymorphs and is, to a large extent, independent of the detrital skeletal opal distribution.

Karst and variations of $CO_2$ concentration during nodule growth, change the carbonate dissolution and silica precipitation rates. The accompanied change of permeability is dependent on the grainsize of the bioclastic carbonate hostrock and defines the density variation in a nodule.

Dense nodules with a thin and porous margin formed in fine-grained siliceous carbonates with carbonate dissolution rates and dissolved silica diffusion rates, that were high relative to silica precipitation rates.

Porous nodules with a dense margin formed in coarse-grained siliceous carbonates with carbonate dissolution rates and dissolved silica diffusion rates that were low relative to silica precipitation rates.

Porous nodules with a dense core formed in clayey carbonates characterised by dissolved silica concentrations, equal to clay mineral saturation concentrations. This could occur, either because a concentration gradient was absent, or because diffusion rates were low and solid-solid conversion became dominant.

Flint nodules can be used to reconstruct depositional and early-diagenetic conditions during the sedimentation of (Tuffaceous) Chalk (Chapter 5), because the flint nodule distribution reflects the early-diagenetic authigenic silica polymorph concentration that was a function of bacterial metabolism in the early diagenetic anoxic redox zones.

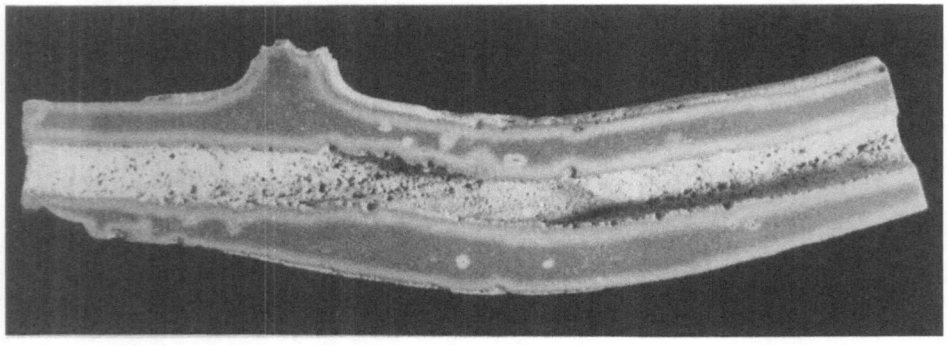

**miscellaneous** - flint concretion around *Thalassinoides* burrow (length 30 cm) in large-scale wavy laminated, bioclastic fine sand of the Emael Member (Maastricht Formation, Maastrichtian s.s., quarry ENCI, Maastricht, The Netherlands). Nodule was found in dissolution residue of karst "sand-pipe" and may have been desilicified at the margins.

# 5  The Genesis of Flint Nodule Layers

## Abstract

**The upper 40 m of the Maastrichtian Gulpen Formation (Maastricht, The Netherlands) is characterised by a regular succession of approximately 75, thickening upwards, laterally continuous flint nodule layers. Flint nodule layers formed when detrital skeletal opal dissolved during late diagenesis and concentrated at sites of relatively high early diagenetic authigenic silica polymorph concentration. Early diagenetic authigenic silica polymorphs precipitated in the anoxic redox zones of bacterial sulphate and/or carbondioxide reduction, that was situated below and parallel to the sediment surface. Highest concentrations of authigenic silica were reached during periods of slowest deposition when sediment resided for a relatively long period in the anoxic redox zone. The rhythmic vertical variation of the flint nodule concentration has been measured with automatic image analysis and reflects the influence of the periodic variation of the Earth's orbital parameters (precession index) on climate, oceanography and periodically varying deposition rates. The relation between the periodic variations of the depositional and early diagenetic conditions and the rhythmically bedded flint nodule layers is simulated with a numerical model. It is shown how the rhythmic variation of authigenic mineral concentrations in Chalk can be used to reconstruct the variation of the dynamics of the shallow marine subtropical Chalk sea.**

## Introduction

The Late Campanian-Late Maastrichtian Gulpen Formation (Felder, 1975a,b; Streel & Bless, 1988) is well exposed in the quarries North, Romont, ENCI and along the incision of the Albert Channel. These outcrops are situated several km apart, west of the River Meuse between the cities of Maastricht (The Netherlands) and Visé

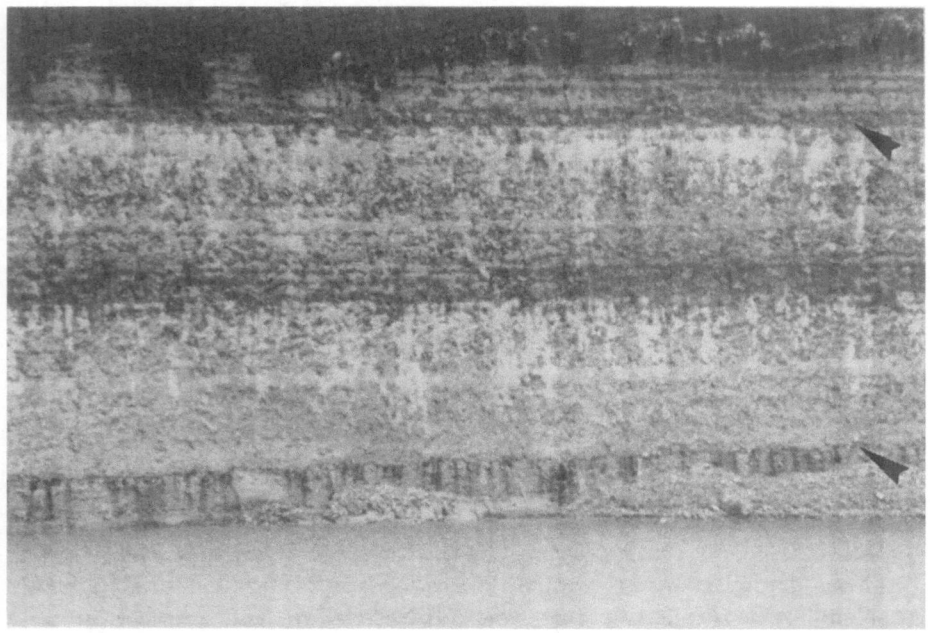

Figure 5.1a - Approximately 9 m thick succession of coccolithic-bioclastic siltstone with planar-parallel bedded flint-nodule layers of the uppermost part of the Lixhe 1 Member, the Lixhe 2 Member, and lowermost part of the Lixhe 3 Member, bounded by the Hallembaye and Boirs Horizons (arrows) (Gulpen Formation, quarry ENCI, Maastricht, The Netherlands).

(Belgium) (Fig. 2.6).

The Maastrichtian part of the Gulpen Formation is a 60 m thick coarsening upward sequence with a superimposed rhythmic variation of the grainsize in dm to m thick beds (Felder, 1986, 1988). The succession consists of homogeneously bioturbated muddy to silty subtropical shallow-marine bioclastic carbonates (Chalk and Tuffaceous Chalk) (Fig. 2.9). The basis of this sequence is smectitic (Vylen Member) and the middle to upper part (Lixhe 1,-2,-3 and Lanaye Members) is characterised by planar-parallel and laterally continuous silica concretion (flint) layers (Fig. 5.1 a,b).

Already in the 19th century, the upper 20 flint layers (Lanaye Member) were used for lithostratigraphical correlation and for the calculation of the tectonic dip (1°to the NW, Binkhorst, 1859; Francken, 1947). More recently, the flint nodule layers of the Gulpen Formation have been recorded in detail and many small exposures of Chalk with Flint, west and east of the River Meuse, could be correlated (Felder, 1975a,b).

The vertical variation of the silica concentration in the Gulpen Formation has

Figure 5.1b - Approximately 20 m thick succession of coccolithic bioclastic siltstone with planar-parallel bedded flint nodule layers of the uppermost part of the Lixhe 3 Member and Lanaye Member, bounded by the Nivelle and Lichtenberg Horizons (arrows) (Gulpen Formation, quarry Romont, Eben, Belgium).

been analyzed quantitatively by means of automatic image analysis. Accurate data of the flint nodule layers are compared to the results of a numerical model that simulates the genesis of the vertical variation of silica concentration as a function of the (periodic) variations of deposition rates and hydrodynamics during the sedimentation.

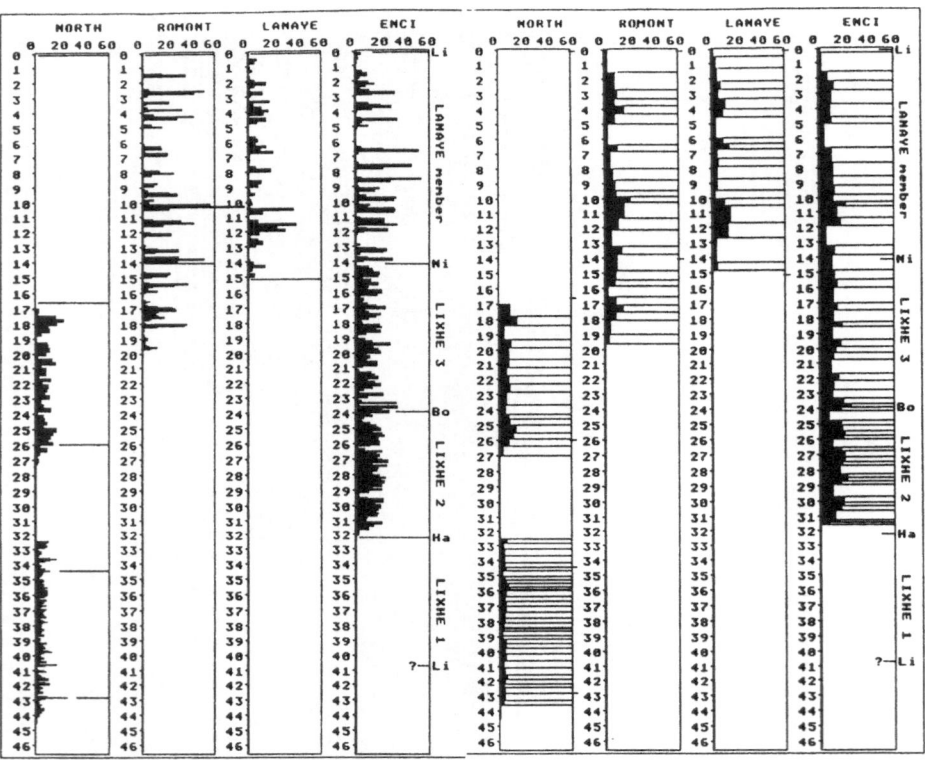

Figure 5.2 (left) - silica nodule concentration (0-80%) measured in 5-20 cm intervals in 45 m of the Gulpen Formation, exposed in the quarries North, Romont, ENCI and the Albert Channel incision (Fig. 2.6). The succession covers the Lixhe 1, Lixhe 2, Lixhe 3 and Lanaye Members, bounded by the Lixhe, Hallembaye, Boirs, Nivelle and Lichtenberg Horizons (Felder, 1975a,b) (for data see appendix B).

Figure 5.3 (right) - average silica concentration (black areas, 0-27%) between successive flint maxima (thin lines). Cycles show thickening upwards and rhythmic variation of the thickness and of the average silica concentration (for data see appendix B).

## 5.1 Quantification of the silica distribution

Clean and smooth vertical quarry walls, with dense (>90% $SiO_2$), bluish-black and grey-coloured cryptocrystalline quartz concretions (Buurman & van der Plas, 1971), easily distinguishable from the grey-white to cream-coloured carbonates, were

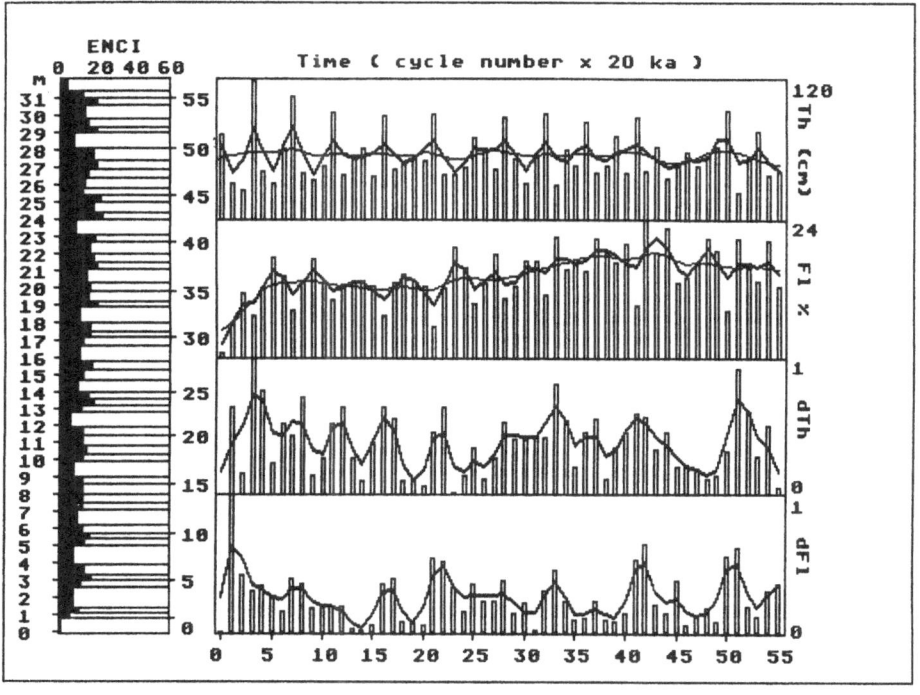

Figure 5.4 - Left: the 32 m thick succession of the upper part of the Gulpen Formation (Hallembaye Horizon=bottom to Lichtenberg Horizon=top) as measured in quarry ENCI (Figs. 5.2, 5.3), and after removal of the thickening upwards trend. The thickness of the ith cycle has been multiplied with the average thickness of the 56 cycles divided by the average thickness of the cycles i-2,i-1,i,i+1,i+2. The average silica concentration (black areas, 0-20%) between the 57 successive flint maxima (thin lines) has been recalculated using the new thickness and the originally measured flint volume between two maxima. Right: four graphs show (from top to bottom) vertical thickness variation (Th, max=120 cm) of the cycles (0-56, old to young, left to right) with 11-12 maxima, as well as the negatively correlated average silica concentration of the cycles (Fl, max=24%) that gradually increases upwards and the variation of the normalized thickness difference (dTh) and silicification difference (dFl) between successive cycles, defined as, i.e. for the thickness, $dTh_{(i)} = ABS(Th_{(i+1)} - Th_{(i)}) / (Th_{(i+1)} + Th_{(i)})$.

photographed at intervals of 2.5 - 10 m hight. The silica concretions visible on large-scale projections of colour slides, were redrawn in black on white paper, in order to guarantee a maximum contrast. A square part of each drawing was recorded with a video camera and the light/dark ratio of 50 successive horizontal bars of 5 - 20 cm thickness in the video image was calculated. The vertical variation of the light-dark ratio of the successive bars is the quantification of the vertical variation of the flint concentration in the (Tuffaceous) Chalk. After scale corrections, the different stratigraphic intervals recorded in the 4 exposures, were

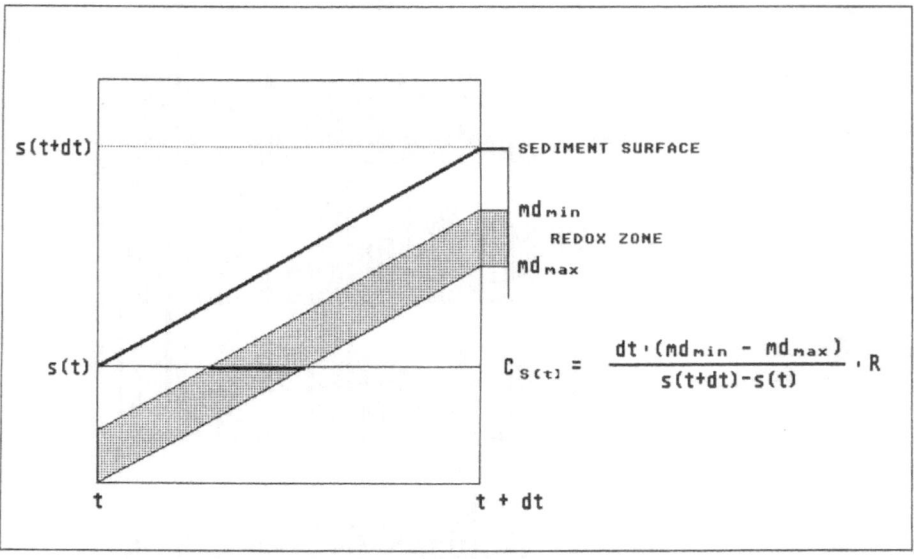

Figure 5.5 - The authigenic mineral concentration ($C_{s(t)}$) that forms in the layer $S(t)$ deposited at t is a function of the deposition/burial rate ($S(t+dt)-S(t)$), the redox zone thickness ($md_{min}$-$md_{max}$) and the mineral precipitation rate (R).

put in order, so that 4 sequences of quantified vertical silica distribution were obtained (Fig. 5.2). The tops of the sequences are formed by the Lichtenberg Horizon and the boundaries between the Members of the Gulpen Formation are indicated (Felder, 1975a,b).

## 5.2  Flint cycles

Local maxima of the recorded silica percentages represent levels of flint nodule layers. Upwards and downwards from a local maximum, the flint percentage decreases. The stratigraphic level of the local minima, situated between successive maxima, is difficult to define, particularly in the Lanaye Member where flint layers are separated by several dm-thick layers without flint. Therefore, a flint cycle has been defined as the interval between two local maxima. The degree of silicification of the cycles has been defined as the average silica concentration between two maxima.

The successions of thus defined flint cycles in the Lanaye Member show a good correlation between 3 of the 4 measured exposures (Fig. 5.3). The most complete and continuous, 32 m thick succession exposed in quarry ENCI, furthermore shows a thickening upward with superimposed a rather regular thickness variation of the flint cycles, in particular in the Lixhe 2 Member. The regularity is accentuated if the thickening upwards trend in the quarry ENCI succession is removed (Fig. 5.4). The thicknesses and the silica concentrations of the cycles, and the difference between thicknesses and silica concentrations of successive cycles show 11-12 maxima in the succession of 57 cycles. The succession can thus be divided into stratigraphic intervals of approximately 5 cycles.

## 5.3 Early diagenetic origin of flint nodule layers

In analogy to the composition of recent siliceous carbonate oozes found in the deep sea (Hurd, 1973; Calvert, 1974; Håkansson et al., 1974; Wise & Weaver, 1974), the (Tuffaceous) Chalk of the Gulpen Formation has been deposited as a watery mixture of fine-grained skeletal carbonate, skeletal opal (Soudry et al., 1981) and organic matter.

During burial, this sediment passed through several zones with different redox conditions, situated at some depth below and parallel to the sea bottom (Berner, 1980). These zones were characterised by specific bacterial metabolism and by authigenic mineral precipitation during the decomposition of organic matter (Redfield et al., 1963; Bischoff & Sayles, 1972; Froelich et al., 1979; Berner, 1980; Zijlstra, 1987,1989; Chapter 3).

In the deepest and most anoxic redox zones sulphate or carbondioxide were reduced and the genesis of hydrogen-sulphides and/or methane (Reeburgh, 1980) led to an increase of the hydroxyl-ion concentration, the dissociation of $H_4SiO_4$ complexes and the precipitation of relatively insoluble silica polymorphs from a pore fluid characterised by relatively high $H_3SiO_4^-$ complexes (Chapters 3, 4).

Assuming that the depth below the sediment surface, the thickness and the reaction rates in the anoxic redox zones have been constant, any vertical variation of the authigenic silica polymorph concentration in the finally buried sediment would have been a function of changing deposition and burial rates only (Fig. 5.5). At high burial rates, sediment passed through the anoxic redox zones relatively fast, and consequently the authigenic silica concentration of the finally buried sediment remained relatively low. To the contrary, during a decrease of the burial rate, the sediments resided longer in the anoxic redox zone, and higher concentrations of authigenic silica were reached. The vertical variation of the early diagenetic authigenic silica concentration was thus inversely proportional to the deposition rate.

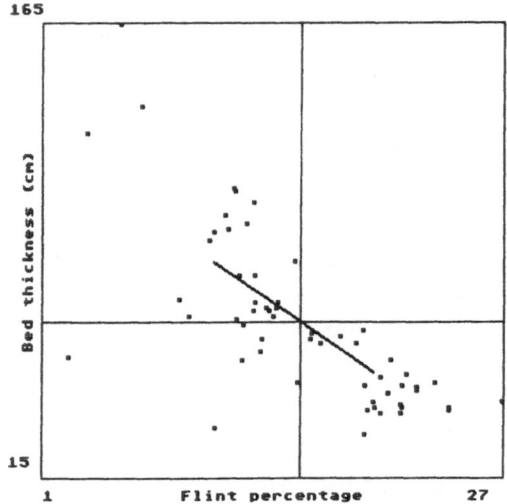

Figure 5.6 - The average silica concentration (1-27%) of a flint cycle is inversely proportional to the flint cycle thickness (15-165 cm). Horizontal line is average bed thickness and vertical line is average flint percentage. Oblique line connects averages of UL and LR quadrants.

## 5.4 Late diagenetic genesis of flint nodule layers

Flint-nodule layers were formed during late diagenesis, when skeletal opal dissolved and monomeric silicic acid diffused towards the levels of the highest early diagenetic authigenic silica polymorph concentration, where it precipitated and caused the growth of flint nodules (Siever, 1962; Williams & Crerar, 1985; Williams et al., 1985; Zijlstra, 1987, 1989; Chapter 4).

It was suggested (Chapter 4) that the morphology, size and concentration of flint nodules is similar to the distribution of the early diagenetic authigenic silica polymorph concentration and hardly depends on the detrital skeletal opal concentration. The silica concentration of a cycle is inversely proportional to the cycle thickness (Fig. 5.6). Considering the presumed constant early diagenetic silicification rate, the inverse proportionality of silica concentration to deposition rate, and the proportionality between early diagenetic silica and flint concentration, then it is suggested that the cycles have been formed during periods of more or less equal duration.

## 5.5 Reconstruction of the depositional environment

Conceivably a proto-flint layer was formed during periods of slow deposition and

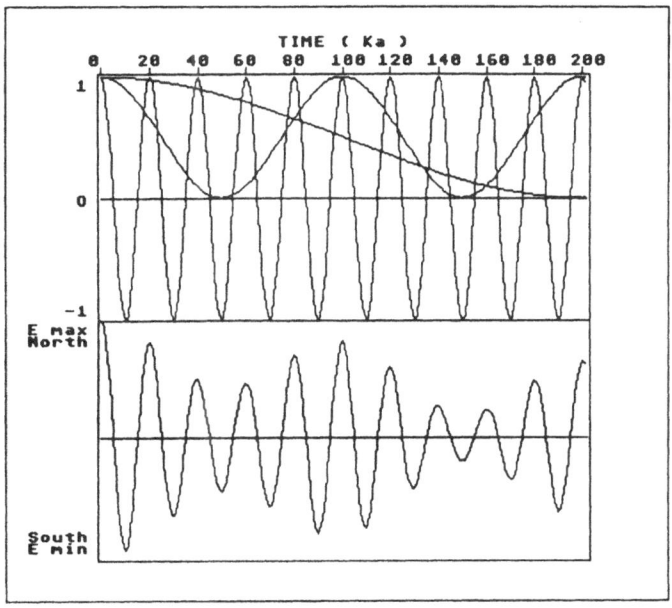

Figure 5.7 - Simplified theoretical curve used for the simulation of orbitally induced periodic changes of the maximum storm intensity.
Sine function with period of 20 ka (Sin1) is multiplied with sine functions with periods of 100 ka (Sin2) and 400 ka (Sin3), using proportionality factors (amplitudes) A1, A2, A3 in order to produce a curve that resembles the periodic north-south shift of the caloric equator and the related periodic variation of the average storm intensity (E(t)=(A1 + A2*Sin2 + A3*Sin3)*Sin1) (E(t)>0).

slow burial, in a sediment layer that remained for a relatively long time in or just below the anoxic zone, at several dm below and parallel to the sediment surface. The upward increase of the thickness of the flint cycles, presumably formed during periods of more or less the same duration, reflects a gradual increase of the average deposition rate, with superimposed short-term periodic fluctuations. The simultaneous increase of the mean grainsize from coccolithic mudstone to coccolithic bioclastic siltstone and the rhythmic variation of the grainsize (Felder, 1986, 1988) indicate that also the hydrodynamic energy of the depositional environment increased while varying periodically.

The deposition rates of the Chalk of NW Europe are estimated to have been of the order of cm-dm/ka (Håkansson et al., 1974, van Hinte, 1976). The dm-m thick flint cycles thus were deposited during several ten thousand years.

The bundling of the flint layers in groups of approximately 5 cycles, the fact that the lithostratigraphic Members coincide with sequences of 20 cycles (Lixhe 1,2 and Lanaye Members) and that the Chalk with Flint of the Gulpen Formation contains approximately 75 beds, strongly suggests the influence of the (quasi-)periodic

76

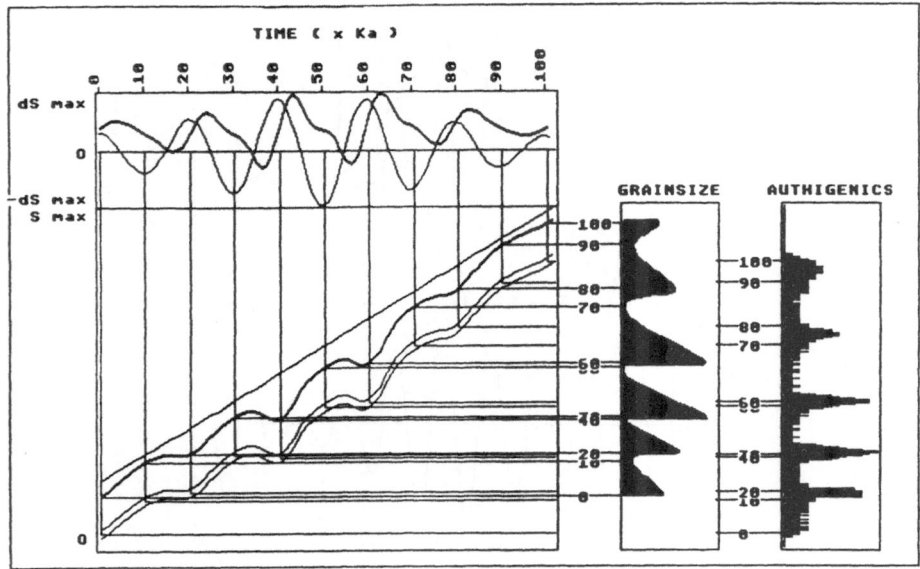

Figure 5.8 - Upper left box: the deposition/erosion rates (fat line) vary periodically between dsmax and -dsmax during 100 ka as a function of subsidence/sea-level rise and storm intensity following the precession index (thin line). Lower left box: during 100 ka sea-level or subsidence remain constant (thin straight line). Water depth and the fair-weather sediment surface (thick wavy line) vary periodically as a function of the storm intensity. A redox zone of mineral authigenesis is situated at constant depth below the fair-weather sediment surface (double thin wavy lines). Vertical-horizontal thin straight lines show the relation between the storm intensity minima and maxima at 10 ka intervals and the grainsize or authigenic mineral concentrations in the sequence 0-Smax (right boxes). Note that the grainsize is proportional to the storm intensity, while the authigenic mineral concentration is inversely proportional to the deposition/erosion rates.

variation of the precession index on sedimentation. The cycles in the Chalk are presumably Milankovitch cycles (Cottle, 1989; Gale, 1989; Hart, 1989; Herrington et al., 1991) and flint cycles formed during, on average, 20 ka precession periods (P) with superimposed modulations due to the variation of ellipticity (P = 20 ka, E1 = 98 ka, E2 = 126 ka, E3 = 413 ka, E4 = 1300 ka; cf. Berger, 1988; Schwarzacher, 1989; Fischer, 1991; De Boer, 1991a,b)

Because the cycles of the Gulpen Formation were presumably deposited during approximately 20 ka precession periods and because pre-Pleistocene short term (third order) sea-level oscillations are thought to have been slow, i.e., of the order of 20 m - 100 m/1-10 Ma (Cooper, 1977; Vail et al., 1977; Hardenbol et al., 1981; Haq et al., 1987), it is suggested that the gradual increase of the deposition rate and the hydrodynamic energy during the deposition of the Gulpen sequence may have been a function of long-term sea-level rise and/or subsidence and that the short-

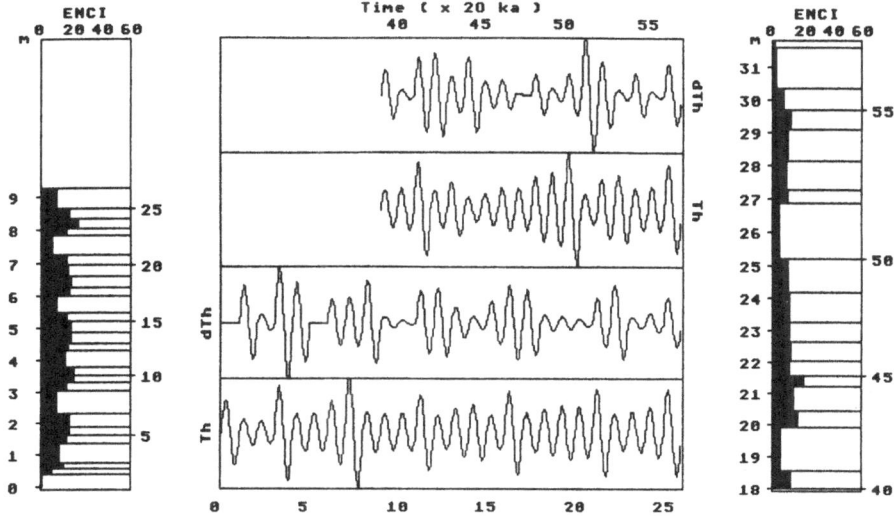

Figure 5.9 - Reconstruction of the orbitally induced storm intensity curve (middle) for the field sequences of the Lixhe 2 (left, 9 m, 25 cycles) and Lanaye Member (right, 14 m, 16 cycles) exposed in quarry ENCI (Fig. 5.2; 5.3). Cycle periods are assumed to have been constant and the intensity (amplitude) has been defined as either being proportional to the bed thickness (Th) or being proportional to the thickness difference of successive beds (dTh). Two lower middle boxes correspond with the Lixhe 2 Member and two upper middle boxes with the Lanaye Member. Record is discontinuous, the Lixhe 3 Member in between is not represented.

term periodic variations of hydrodynamic energy and deposition rate were a function of short-term periodic variations of climate and oceanography. In particular the change of average storm frequency and intensity might have caused the variation of the hydrodynamic energy and deposition rates (Birdsall & Steward, 1978; Aigner, 1979, 1982; Wright et al, 1986; Marsaglia & Klein, 1983), which might have been related to e.g. periodic latitudinal shifts of the caloric equator (cf. Berger, 1979; de Boer, 1983, 1991a,b) (Fig. 5.7).

## 5.6 Numerical simulation of flint nodule layer genesis

In order to understand the relation between the variation of hydrodynamics and the deposition rate and the variation of the authigenic silica concentration in the Chalk, a numerical model has been developed (appendix A). In the model, the deposition rate varies periodically, in phase with the orbitally induced variations of storm

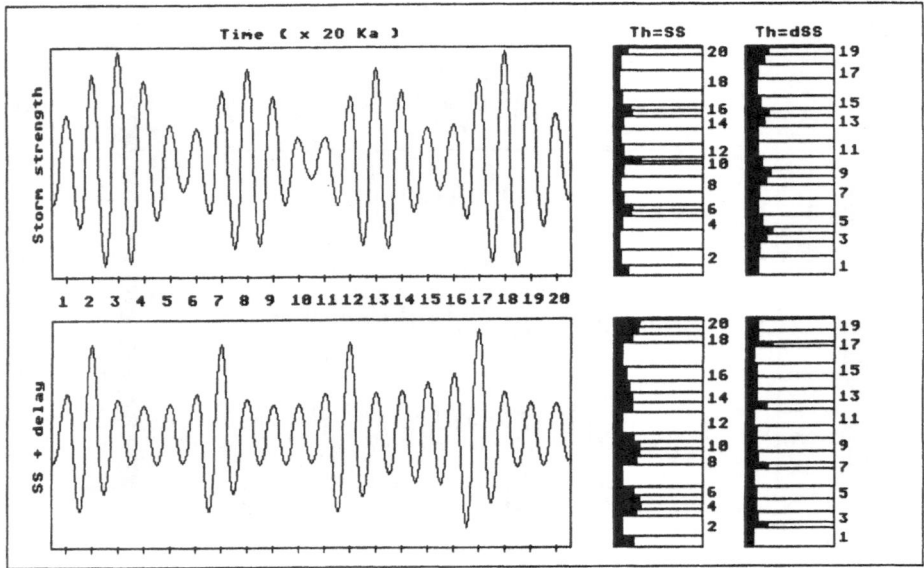

Figure 5.10 - Upper left box: reconstruction of flint cycle succession using a theoretic storm intensity (SS) curve and, lower left box: the same curve (SS + delay) with an implemented delay. Flint cycle successions are calculated with cycle thickness proportional to the amplitude (Th=SS) (upper box for SS curve and lower box for SS+delay curve) and with cycle thickness proportional to the difference between successive amplitudes (Th=dSS) (upper box for SS curve and lower box for SS+delay curve). Note that the sequences produced by the SS + delay curve (lower right boxes) are most similar to the field sequences (Fig. 5.9), and that the Th=SS sequence resembles the Lixhe 2 cycle succession, and the Th=dSS sequence resembles the Lanaye cycle succession.

intensity, and reaches a relative maximum during each precession period (Fig. 5.8). During a precession period, the basin is thus characterised by a time of high hydro-dynamic energy when storm intensity is at a maximum, preceeded and followed by times of minimum hydrodynamic energy, when storm intensity is at a minimum.

During the increase of the hydrodynamic energy, sediment was eroded and carried out of the environment of deposition. While the subsurface subsided and the water depth increased, net deposition rate decreased and a flint layer started to form. During the decrease of the hydrodynamic energy, sediment accumulated again. While the deposition rate increased and the water depth decreased, a relatively authigenic silica- poor layer was deposited. The magnitude of the periodic variation of the deposition rate was a function of the difference between maximum and minimum storm intensity during a precession period and of the long-term change of average hydrodynamic energy of the depositional environment that depended on the gradually changing basin morphology.

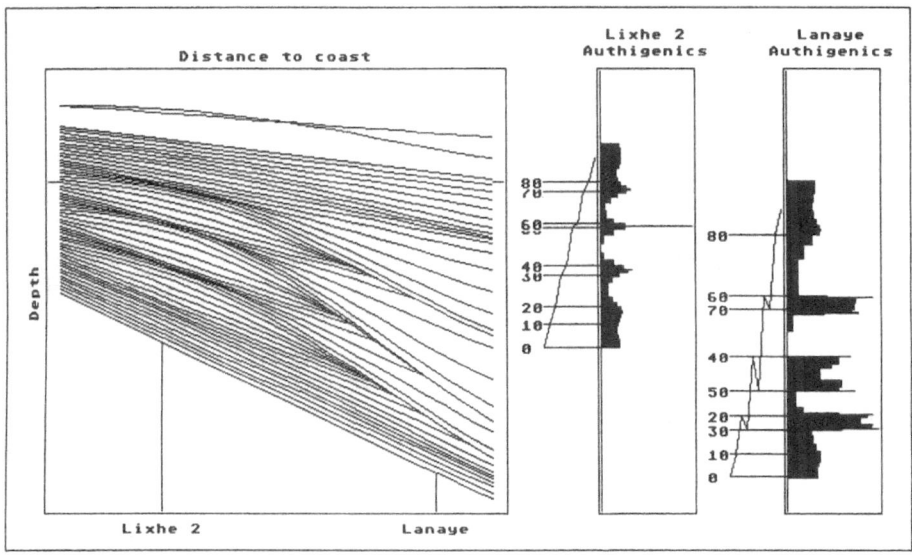

Figure 5.11 - Explanation of the observed differences between cycle successions of the Lixhe 2 and Lanaye Member. Left box: The seabottom changes as a function of the storm frequency-intensity that follows the precession index. During low energy periods the seabottom is a planar surface with gentle dip. During high energy periods the seabottom is a curved surface with locally steeper dip (from coast left to off-shore right). The redox zone of silica authigenesis (not depicted) is situated at constant depth below and parallel to the seabottom. The form and position of the seabottom has been calculated for a 100 ka period (2 ka intervals) with periodically varying storm intensity (Fig. 5.8), while the subsurface rotated and subsided. In the near-coastal realm thin-bedded symmetric cycles are formed in fine-grained Chalk (Lixhe 2 Member), while in the more open marine realm thin- and thick-bedded asymmetric cycles are formed in coarser-grained Tuffaceous Chalk (Lanaye Member). Authigenic mineral concentrations and the position of the redox zone at 10 ka intervals for symmetric (Lixhe 2) and asymmetric (Lanaye) sequences are depicted in the right boxes.

According to the model, the lowest deposition rates, during the genesis of an eccentricity cycle, occurred when the maximum storm intensity of successive precession storm periods increased during half the eccentricity period, while during the other half the reversed occurred.

## 5.6.1 Simulation results

The simulation of the genesis of authigenic silica layers during a 100 ka eccentricity period, assuming constant subsidence rate and a periodic variation of the deposition rate as a function of a simplified storm intensity curve, comparable to the calculated latitudinal shifts of the caloric equator (Berger, 1979), produces an asymmetric succession of thickening-upward cycles, with authigenic mineral concentration inversely proportionally to the cycle thickness (Fig. 5.8). However, if the field successions (Lixhe 2 Member, Lanaye Member) are considered the output of the simulation model and if the storm intensity is reconstructed, using the model in reverse, then it appears that the reconstructed curve differs considerably from the initially presumed theoretical curve (Fig. 5.9).

The reconstructed storm-intensity curve is asymmetrical and suggests that the increase-decrease of hydrodynamic energy and the decrease-increase of deposition rate lag behind the increase-decrease of the storm intensity. This means that the sedimentation rate did not change in phase with the change of storm intensity, in particular when the rate of storm intensity change was at a maximum. Another cause for the lag might be the syn-depositional lithification and the consequent decrease of maximum erosion depth (pers. comm. Schuiling; Chapter 6).

Indeed, if the initial theoretical storm intensity curve is modified by implementing a lag and transformed into a hydrodynamic energy curve, then, with the model, sequences can be generated that are much more similar to the observed field sequences (Fig. 5.10). Furthermore, one may distinguish a rather symmetrical cycle succession of which the cycle thicknesses are proportional to the storm intensity maxima of the subsequent precession periods (SS), and an asymmetrical cycle succession, of which the cycle thicknesses are proportional to the differences between the storm intensity maxima during successive precession periods (dSS). The first eccentricity cycle type is comparable to those of the Lixhe 2 Member and the second is comparable to those of the Lanaye Member (Figs. 5.9, 5.10). Apparently the cycles of the Lixhe 2 Member reflect an environment of mainly deposition, although varying, while the cycles of the Lanaye Member reflect times of considerable erosion during successive precession periods.

If one assumes that the fine-grained Chalk of the Lixhe 2 Member with symmetric cycles has been deposited on a mudflat and that the coarser-grained asymmetric (Tuffaceous) Chalk cycles of the Lanaye Member have been deposited in a more energetic open marine realm, then the genesis of both cycle types can be simulated, when the seabottom morphology varies between, planar gently dipping during times of low storm intensity, and curved, locally steeper, during times of higher storm intensity, while simultaneously, long-term change of the basin morphology is characterised by subsurface rotation and subsidence (Fig. 5.11).

## Discussion

Since the recognition of orbital cycles as the cause of variations in the intensity and distribution of solar insolation and the hypothesis that these variations caused periodic glaciation during the Late Tertiary, numerous regularly bedded sedimentary sequences of all ages have been investigated and were shown to have preserved the effect of the orbital variations as a rhythmic vertical variation of the bed thickness and/or sediment composition (Hays et.al, 1976; Imbrie & Imbrie, 1979).

Proving the Milankovitch rhythmicity requires ideal sequences, deposited at constant rate, over a long period and with a variation of sediment composition that is proportional to the variation of the intensity and distribution of insolation. Most attention has therefore been focused on such sequences (cf. Fischer, 1991).

The non-ideal rhythmic successions of the Chalk with Flint of the Gulpen Formation have been subdivided into 4 lithostratigraphic members at a time when Milankovitch cycles had not yet been recognized (Uhlenbroeck, 1912). Therefore, the fact that the Chalk with Flint contains 75 flint layers, that the members contain respectively 20, 15, 20 and 20 flint layers and that the flint layers form bundles of 5 layers, at least strongly suggests the presence of respectively E4 (1300 ka), E3 (413 ka), E2 and E1 (98 and 126 ka) eccentricity cycles and P (20 ka) precession cycles (cf. Fisher, 1991). Above I have presumed the presence of Milankovitch cycles, because the inability to prove their existence does not entail their absence, and I have presented a model where orbitally induced sedimentary cycles are used as a clock with which deposition rates can be defined more accurately. It has thus been shown how in non-ideal sequences, the way in which the influence of orbital variations has been preserved, provides information about the dynamics of the depositional-early diagenetic environment, characterised by the variation of hydrodynamic energy and of deposition rate and by the variation of seabottom morphology during sub-surface motion and/or sea-level change.

# Conclusion

In fine-grained and skeletal opal-rich Chalk, authigenic silica formed at a presumed constant rate in anoxic redox zones of bacterial metabolism with a presumed constant thickness, several decimetres below and parallel to the sediment surface. At times of relatively low deposition and burial rates, sediments resided longer in the redox zones of silica precipitation and consequently levels of elevated concentration were formed. Early diagenetic silica concentration was inversely proportional to the deposition rate. During late diagenesis, detrital skeletal opal further dissolved and silica diffused towards and precipitated at the levels of highest early diagenetic silica polymorph concentration, thus forming a flint nodule layer.

Sequences of Chalk with flint-nodule layers reflect a periodic variation of the deposition rate. The periodic variation of the deposition rates was a function of the periodic variation of the Earth's orbital parameters, climate and oceanography. Flint cycles have been formed during approximately 20,000 year long precession periods.

The 40 m thick succession of the Late Maastrichtian Gulpen Formation, exposed in quarries near Maastricht (The Netherlands), contains about 75 precession induced sedimentary cycles and thus would have been deposited during approximately 75 x 20 = 1.5 million years. The increase of the mean cycle thickness from 4 dm at the basis to 1 m at the top of the sequence then reflects an increase of the average deposition rate from 2 to 5 cm/ka.

Fine-grained Chalk with a high flint concentration, and with symmetrical thin-bedded eccentricity cycles reflects a low energy environment with a low deposition rate. Coarser-grained (Tuffaceous) Chalk with a lower flint concentration and with asymmetrical thick-bedded eccentricity cycles reflects a high energy environment with high erosion/deposition rate.

Besides the classical tools, i.e., the analysis of (trace) fossils and grain-size distribution, also authigenic mineral distribution and bed-thickness variations provide, due to the effects of orbital variations, information about the character of the depositional and early diagenetic environment.

## Appendix A - Numerical model

The relation between authigenic mineral concentration and deposition rate has been investigated analytically for sedimentary environments that were characterised by variable but positive deposition rates (deposition only) (Berner, 1980). For environments that are characterised by erosion, a numerical model is better suited for the simulation of the vertical variation of authigenic mineral concentration as a function of sedimentation rates.

The numerical model requires the definition of the variation of the deposition-erosion rates for a period t=0 to t=T. The deposition/erosion rates are a function of subsidence and/or sea-level rise (procedure 1) and of the storm frequency-intensity (procedure 2). The storm frequency-intensity is presumed to vary periodically between 0 and 1 with periods of 20, 100 and 400 ka and the effect of the storm frequency-intensity on the deposition rates is a function of the hydrodynamic properties of the basin (procedure 3).

After definition of the deposition/erosion rates during a period T, and with steps dt a numerical array is filled (deposition) with a sediment that has a grainsize proportional to the hydrodynamic energy, or it is emptied (erosion). Deposition or erosion occur at discrete steps until a sequence of S layers has been formed. During each period dt, after sediment has been deposited or eroded, authigenic silica concentration is increased in a zone of one or more layers thickness at constant depth below the sediment surface and characterised by a constant reaction rate (procedure 4).

```
+ = addition
- = substraction
* = multiplication
/ = division
m^n = m to the power n
SQR(n) = square root of n
A(n) = array of n sites
pi = 3.14..
```

```
GO TO PROCEDURE 1 (average deposition rate)
GO TO PROCEDURE 2 (periodic variation of hydrodynamic energy)
GO TO PROCEDURE 3 (periodic variation of the deposition rate)
GO TO PROCEDURE 4 (grainsize and authigenic mineral concentration)
```

PROCEDURE 1 - The definition of the deposition rate and the position of the sediment surface in time as a function of the rate of subsidence and/or sea-level rise.

```
FOR t=1 TO t=T
        dS(t)=dSmin+(dSmax-dSmin)*(t/T)
        S_tot=S_tot+dS(t)
NEXT t
```

```
FOR t=1 TO t=T
        dS(t)=dS(t)*((S-S0)/S_tot)
NEXT t
FOR t=0 TO t=T
        IF t=0
                        S(t)=S0
        ELSE
                        dS1=dS(t)+dS3
                        dS2=INT(dS1+0.5)  INT( )=integer
                        dS3=dS1-dS2
                        S(t)=S(t-1)+dS2
        ENDIF
NEXT t

dS(t) = deposition rate at t for t=0 to t=T
S(t) = sediment surface at t for t=0 to t=T
S0 = sediment surface at t=0
S = sediment surface at t=T; S>S0
dSmin = minimum deposition rate
dSmax = maximum deposition rate
S_tot = 0
RETURN
```

PROCEDURE 2 - The definition of the variation of the theoretical storm intensity in time resembles the latitudinal shift of the caloric equator (cf. Berger, 1979).

```
FOR t=0 TO t=T
        E1=sin(((2*pi)/P1)*t)
        dt=ph*P2*(sin((2*pi/P2)*t-pi/2))  dt=Lag (0<ph<<1)
        E2=(0.5+0.5*sin(((2*pi)/P2)*(t+dt)))
        E3=(0.5+0.5*sin(((2*pi)/P3)*t))
        E(t)=0.5*(1+(A1+A2*E2+A3*E3)*E1)
NEXT t

E(t) = storm intensity at t from t=0 to t=T (0<E(t)<1)
A1 = 0.15 (20 ka precession amplitude)
A2 = 0.6*(1-A1) (100 ka eccentricity amplitude)
A3 = 0.4*(1-A1) (400 ka eccentricity amplitude)
P1 = 20 (precession period)
P2, P3 = 100, 400 (eccentricity periods)
RETURN
```

PROCEDURE 3 - The definition of the change of the position of the sediment surface in time, as a function of the storm intensity and basin form.

```
FOR t=0 TO t=T
        B(t)=Bmin+(Bmax-Bmin)*(t/T)
NEXT t
FOR t=0 TO t=T
        S(t)=S(t)-(E(t)*B(t))
NEXT t
```

B(t) = basin hydrodynamics at t during t=0 to t=T 0<B(t)<S0
Bmin = minimum hydrodynamic energy
Bmax = maximum hydrodynamic energy
RETURN


PROCEDURE 4 - The definition of the grainsize (G(s)) as a function of storm intensity and the concentration of authigenic silica (Si(s)) in the anoxic redox zones at some depth below the sediment surface.

```
FOR t=1 TO t=T
      IF S(t+1)>S(t)
                  FOR s=S(t) TO s=S(t+1)
                            G(s)=f*E(t)*B(t)
                  NEXT s
      ELSE
                  FOR s=S(t+1) TO s=S(t)
                        G(s)=0
                        Si(s)=0
                  NEXT s
      ENDIF
      FOR s=S(t)-RZmax TO s=S(t)-RZmin
            Si(s)=Si(s)+R
      NEXT s
NEXT t
```

G(s) = grainsize of layer s from S0 to S
f = proportionality factor
RZmin = minimum depth of redox zone below sediment surface
RZmax = maximum depth of redox zone below sediment surface
Si(s) = authigenic silica of layer s from S0-RZmax to S-RZmin
R = reaction rate of authigenic silica genesis
RETURN

**Appendix B - Data** of the vertical variation of the silica concentration in the quarries North, Romont and ENCI and the Lanaye channel incision, like depicted in figures 5.2, 5.3.

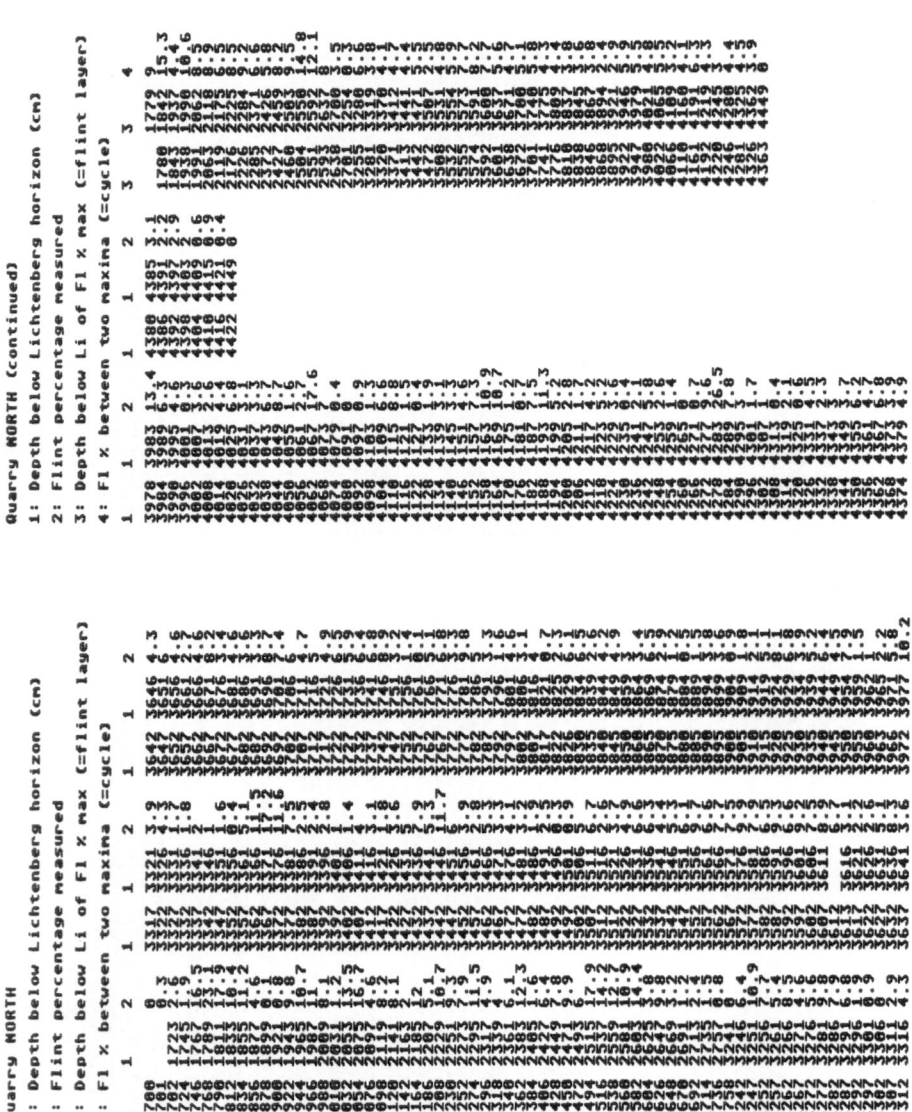

Quarry NORTH (continued)

1: Depth below Lichtenberg horizon (cm)

2: Flint percentage measured

3: Depth below Li of Fl X max (=flint layer)

4: Fl X between two maxima (=cycle)

Quarry NORTH

1: Depth below Lichtenberg horizon (cm)

2: Flint percentage measured

3: Depth below Li of Fl X max (=flint layer)

4: Fl X between two maxima (=cycle)

Albert channel incision LANAVE

1: Depth below Lichtenberg horizon (cm)
2: Flint percentage measured
3: Depth below Li of Fl X max (=flint layer)
4: Fl X between two maxima (=cycle)

Quarry ROMONT

1: Depth below Lichtenberg horizon (cm)
2: Flint percentage measured
3: Depth below Li of Fl X max (=flint layer)
4: Fl X between two maxima (=cycle)

Quarry ENCI

1: Depth below Lichtenberg horizon (cm)
2: Flint percentage measured
3: Depth below Li of Fl X max (=flint layer)
4: Fl X between two maxima (=cycle)

Quarry ENCI (continued)

1: Depth below Lichtenberg horizon (cm)
2: Flint percentage measured
3: Depth below Li of Fl X max (=flint layer)
4: Fl X between two maxima (=cycle)

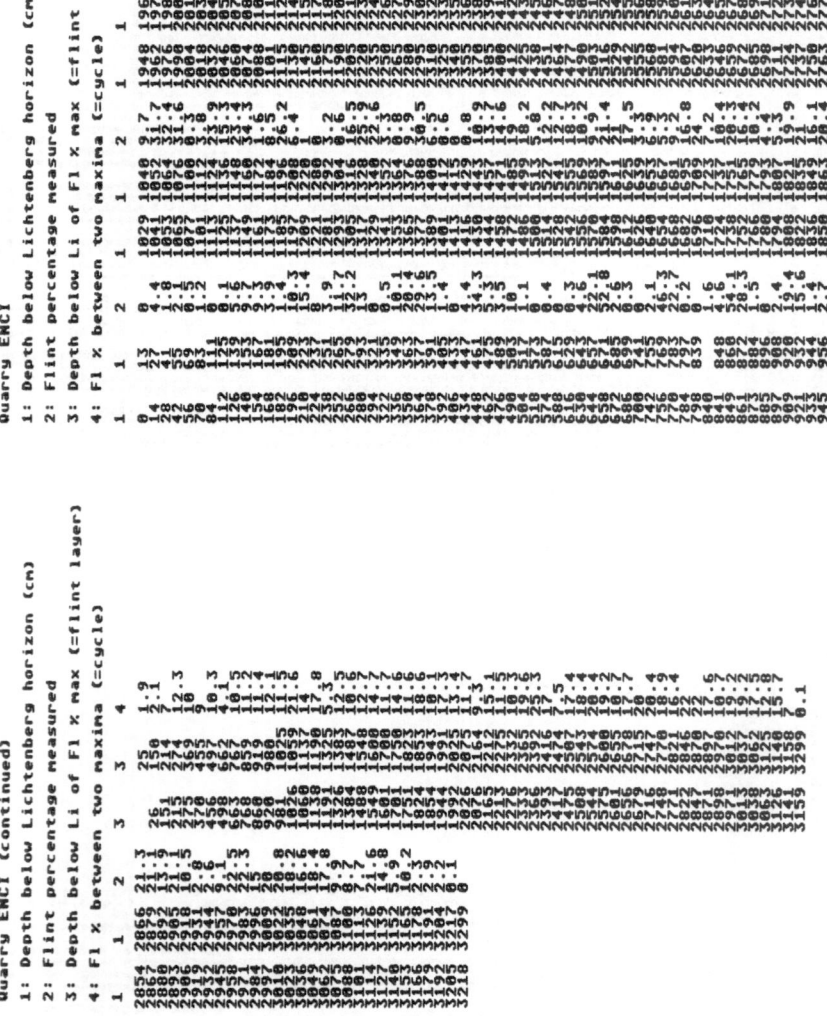

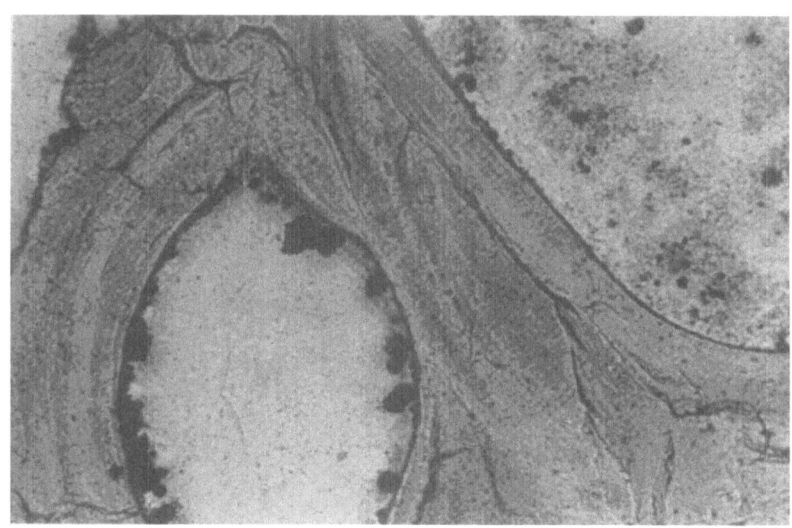

**miscellaneous** - thin section (16x, normal and polarized light (below)) of silicified bone fragment (Gulpen Formation, Maastrichtian, quarry Romont, Eben, Belgium). The 2.5 mm wide vacuole is covered by pyrite crystals at the margins and has been filled with silica growing from the margins to the centre. Note "granular" silicification of sediment in upper right-hand corner.

# 6 The Genesis of Tempestite Cycles

### Abstract

**At the boundary between the Late Maastrichtian Gulpen and Maastricht Formations, exposed in quarry ENCI (Maastricht, The Netherlands), rhythmically bedded coccolithic bioclastic siltstones with flint nodule layers (Chalk with Flint) are covered by rhythmically bedded coccolithic, phosphatic-glauconitic-pyritic bioclastic sandstones with carbonate cement (Tuffaceous Chalk). The rhythmic vertical variations of the grain size, depositional/bioturbational structures and authigenic mineral concentrations in Chalk with Flint and Tuffaceous Chalk with "proto-hardgrounds" were caused by the periodic variation of hydrodynamics and deposition rates. A numerical model enables the simulation of the genesis of rhythmically bedded (Tuffaceous) Chalk sequences as a function of deposition rates, bioturbative mixing, storm reworking and mineral authigenesis in redox zones of bacterial metabolism. The numerical model facilities the discussion of the relation between the variation of the lithology of (Tuffaceous) Chalk sequences and the variation of the dynamics of the depositional-early diagenetic environment in the subtropical shallow marine (Tuffaceous) Chalk sea.**

## Introduction

The Maastrichtian type section of Dumont (1849) is exposed in quarry ENCI (Maastricht, The Netherlands, Fig. 2.6). It is equivalent with the Maastricht Formation (Felder, 1975a,b) and forms the upper half of a 100 m thick carbonate sequence. The lower half of the sequence belongs to the upper part of the Gulpen Formation (Felder, 1975a,b). Today, the upper part of the Gulpen Formation and the Maastricht Formation are both considered to be Late Maastrichtian in age (Jeletzky, 1951; Schmid, 1959; Romein, 1962, 1963; Felder et al, 1985; Streel &

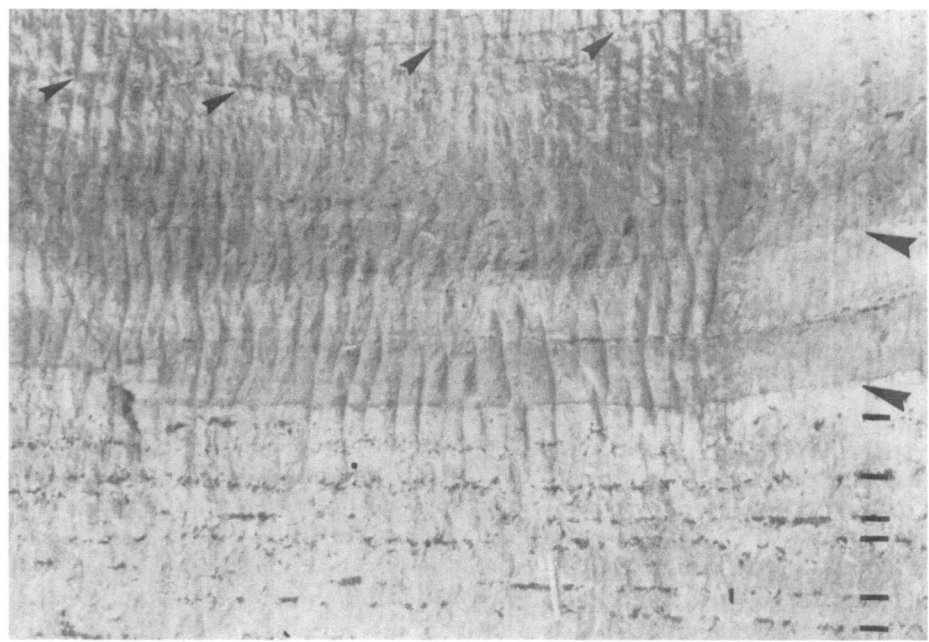

Figure 6.1a - About 5 m of Chalk of the Lanaye Member with flint nodule layers 14-19 (stripes), ending with the Lichtenberg Horizon (large arrow), covered by about 2.5 m of rhythmically bedded phosphatic-glauconitic and pyritic Tuffaceous Chalk of the Valkenburg Member, ending with the St. Pieter Horizon (large arrow), covered by about 5 m of the Gronsveld and Emeal Members. Oblique-bedded, laterally restricted flint layers occur in the upper part of the sequence (arrows) (NE of quarry ENCI).

Bless, 1988).

The Late Maastrichtian sequence gradually coarsens upwards from smectitic coccolithic mudstone (Chalk, Vijlen Member, Gulpen Formation) via homogeneously bioturbated coccolithic bioclastic siltstone with flint nodule layers (Lixhe 1,2,3 and Lanaye Members, Gulpen Formation) towards coarse-grained laminated phosphatic-glauconitic-pyritic bioclastic sand with hardgrounds (Tuffaceous Chalk, Valkenburg to Meerssen Members, Maastricht Formation) (Fig. 2.9).

The sequence is rhythmically bedded. The planar, parallel beds are laterally continuous over distances of kilometres and they all dip gently towards the North-west (<1°; Francken, 1947). The beds gradually thicken upwards from 0.25 m at the basis to more than 2 m at the top of the sequence. Bedding is defined by a rhythmic variation of the grainsize (Felder, 1986), of depositional/bioturbational structures and of the concentrations of glauconite, pyrite, carbonate cement

Figure 6.1b - About 6 m of (Tuffaceous) Chalk with the uppermost part of the Lanaye Member (flint layer 23 (20)), ending with the Lichtenberg Horizon (arrow), covered by 3 phosphatic-glauconitic-pyritic cycles of the Valkenburg member, ending with a poorly developed "proto" hardground below the St. Pieter Horizon (arrow) (Fig. 6.5) and at least 3 phosphatic-glauconitic-pyritic cycles of the Gronsveld Member (quarry ENCI, centre).

(hardgrounds) and silica (flint). It was shown (Bromley et al., 1975; Clayton, 1986; Zijlstra, 1989; Chapter 3) that these minerals are authigenic and formed during early diagenesis in aerobic to anoxic redox zones of bacterial metabolism, e.g. around deep burrows and below and parallel to the seabottom.

For instance, the planar, parallel-bedded flint nodule layers of the Gulpen Formation occur at levels of highest early diagenetic authigenic silica concentration. These formed when, during periods of slow deposition and burial, sediments resided for a relatively long time in the anoxic redox zones of sulphate and/or carbondioxide reduction, several dm below and parallel to the sediment surface, where authigenic silica precipitated. The authigenic silica concentration was inversely proportional to the deposition/burial rate, and the rhythmic succession of flint nodule layers thus reflects a periodic variation of the deposition rate (Chapter 5).

In a several metres thick succession below and above the boundary between the Gulpen and Maastricht Formations (Lichtenberg Horizon) exposed in quarry ENCI

94

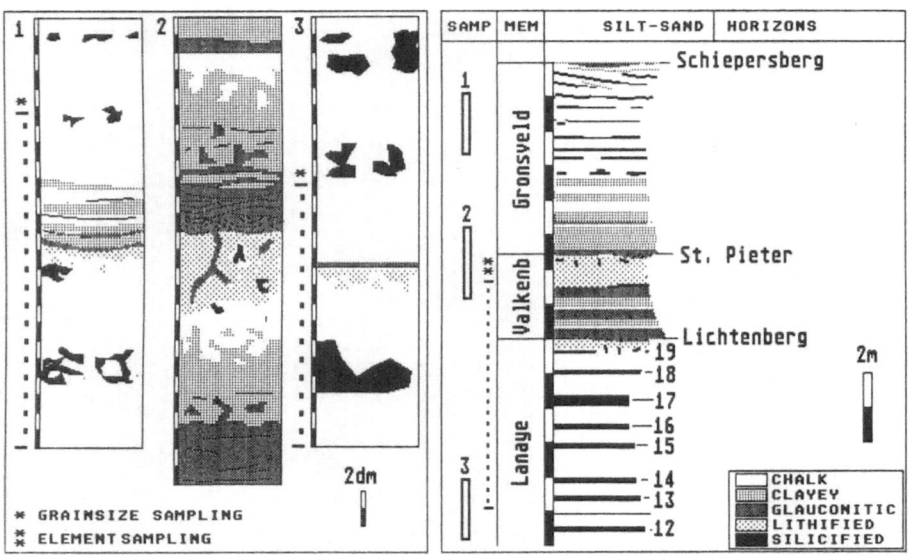

Figure 6.2 - Lithostratigraphy of the boundary zone between the Late Maastrichtian Gulpen and Maastricht Formations (Maastrichtian *s.s.* Dumont, 1849) exposed in quarry ENCI (Fig. 6.1a). Detailed sections 1 (Fig. 4.2) ,2 and 3 show cycles with erosion surfaces of respectively the Gronsveld, Valkenburg/Gronsveld and Lanaye Members. Detailed sections 1 and 3 have been sampled for grain-size analysis (Fig. 6.8). The Lanaye and Valkenburg Members have been sampled for element analysis and microfacies (Fig. 6.7a).

(Fig. 6.1a,b), cycles of homogeneously bioturbated Chalk with flint nodule layers (Lanaye Member) are succeeded by laminated, phosphatic-glauconitic-pyritic fining-upwards Tuffaceous Chalk cycles with poorly developed and laterally restricted flint nodule layers (Valkenburg and Gronsveld Member).

The boundary sequence has been sampled at dm intervals and samples were analyzed for element concentrations (ICP), grainsize (laser particle sizer) and micro facies (thin sections). The results show a rhythmic vertical variation of structures, grainsize, microfacies and of the concentration of authigenic minerals.

It is suggested that the cycles of the Tuffaceous Chalk reflect, like the flint nodule layers in the Chalk (Chapter 5), a periodic variation of hydrodynamics and deposition rates. A numerical model is presented that relates the rhythmic variation of structures, grain size and authigenic mineral concentrations in the Maastrichtian (Tuffaceous) Chalk sequences to the rate of relative sea-level variations, to the periodic variations of the hydrodynamic energy and to the periodic variations of the rate of deposition in the Maastrichtian subtropical shallow marine environment at Maastricht.

Figure 6.3 - Polished slab (approx. 60 cm) of the boundary (Lichtenberg Horizon, large arrow) between the Lanaye and Valkenburg Members. The light-coloured, homogeneously bioturbated muddy bioclastic silt of the Lanaye Member has a wavy erosion surface on top and has been penetrated by *Thalassinoides* burrows filled with coarse glauconitic-pyritic bioclastic sand. The Lichtenberg Horizon is covered by phosphatic-glauconitic-pyritic bioclastic sand that forms the basis of a fining-upwards cycle, which is covered by a second fining upwards cycle of the Valkenburg Member (small arrow). Depositional lamination has been almost entirely destroyed by bioturbation.

# 6.1 Litho-/biofacies

### 6.1.1 Lanaye Member

The Chalk of the Lanaye Member forms the top of the Gulpen Formation and is a 97% pure coccolithic bioclastic silty packstone. The white Chalk has been homo-geneously bioturbated and only locally large-scale wavy lamination has been preserved. Well developed planar-parallel flint nodule layers succeed each other at

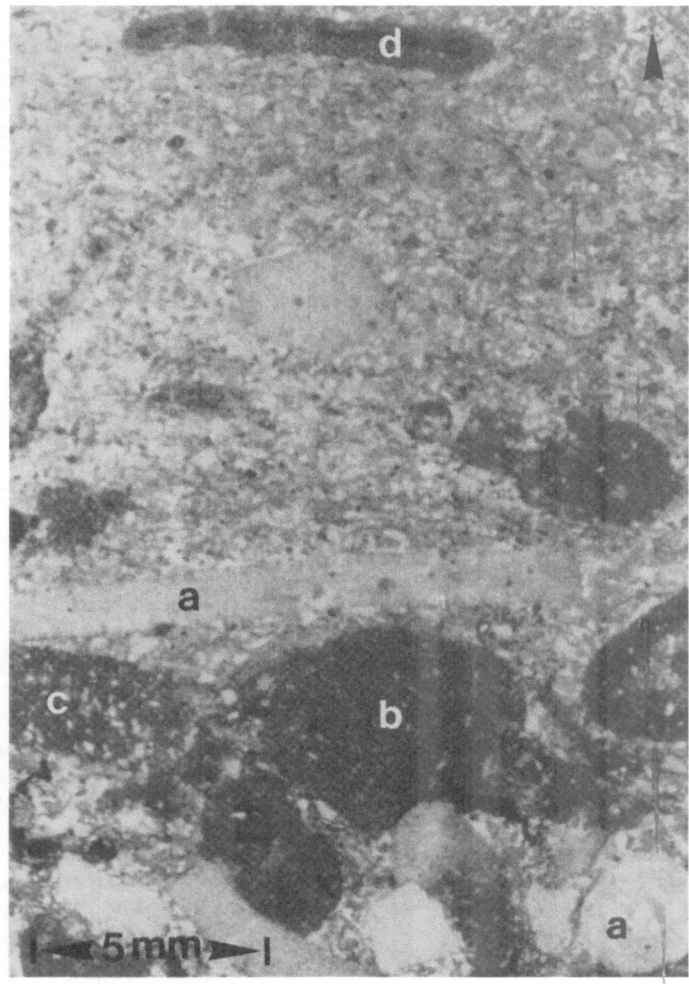

Figure 6.4 - Thin section of the coarse-grained compacted phosphatic-glauconitic-pyritic bioclastic
sand just above the Lichtenberg Horizon. Bioclasts (a), lithoclasts of reworked muddy bioclastic
silt (b) (Lanaye Member), coprolithes (c) and mineralised *Chondrites* burrow fill (d) float in a
matrix of fine-grained bioclastic sand (6x).

0.5 -1.5 m intervals. Part of the nodules are tubular and formed around
*Thalassinoides* burrows (layer 19 of Felder, 1975 a,b; Figs. 6.1a, 6.2), while other
nodules are platy (layer 17). Platy nodules may be oversilicified tubular nodule
layers or may have formed independently of bioturbation structures within
laminated sediment.

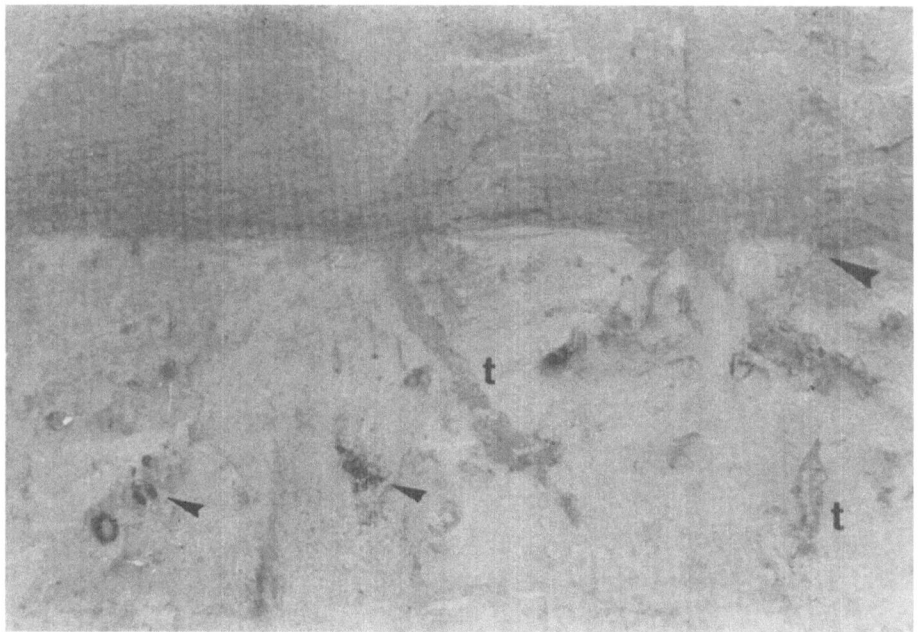

Figure 6.5 - About 0.75 m high sequence forming the boundary between the Valkenburg and Gronsveld Members (St Pieter Horizon, large arrow). Light-coloured and slightly lithified top of a cycle (proto hardground) with poorly developed flint nodules around spreiten burrows (small arrows) and *Thalassinoides* (t) has been eroded and covered by a coarse-grained laminated glauconitic-pyritic bioclastic sand that fines upwards and develops into a slightly lithified and homogeneously bioturbated carbonate again. *Thallassinoides* burrows filled with coarse-grained glauconitic-pyritic sand penetrate the top of the Valkenburg Member (Fig. 6.1b).

Macrofossils are rare in the Chalk of the Lanaye Member. Traces of shallow bioturbation and sediment mixing are common but poorly preserved. Trace fossils of deep burrowers (e.g., *Chondrites, Thalassinoides*) have been preserved as ghost structures in flint (Bromley & Ekdale, 1986).

## 6.1.2. Valkenburg Member

The Tuffaceous Chalk of the Valkenburg Member forms the basis of the Maastricht Formation. The boundary between the Lanaye Member and the Valkenburg Member consists of an irregular, slightly undulating erosion surface (Lichtenberg Horizon). Several tens of metres wide and dm deep depressions at the top of the Chalk of the Lanaye Member are filled with coarse-grained phosphatic-glauconitic-

98

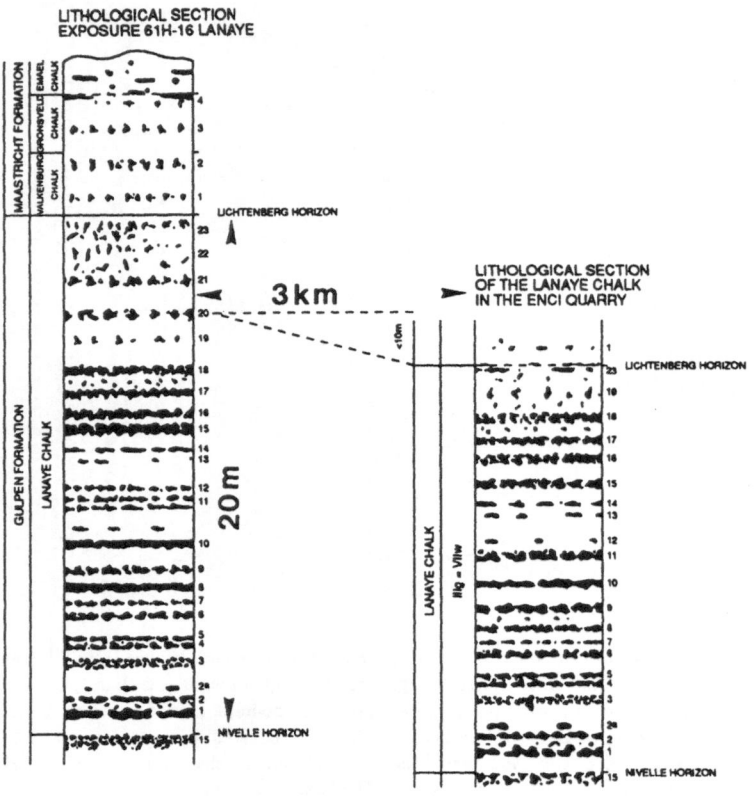

Figure 6.6 - Lanaye Member as exposed at Lanaye (20 m) and ENCI, 3 km further to the North and less than 10 m lower (after Felder, 1975a). Note that flint layer 23 below the Lichtenberg Horizon as defined in quarry ENCI, is chronostratigraphically equivalent with flint layer 20 of the Lanaye Chalk exposed at Lanaye. The lithostratigraphically equivalent Lichtenberg Horizon at Lanaye is most probably chronostratigraphically equivalent to the St. Pieter Horizon in quarry ENCI (Figs. 6.2, 6.5).

pyritic bioclastic sand (Fig. 6.3). The bioclastic sand consists of broken and rounded skeletal remains (echinoids, crinoids, oysters, belemnites, fish and shark teeth), coprolithes and clasts of reworked Chalk (Fig. 6.4). Sand-sized extrabasinal quartz and heavy mineral grains (van Harten, 1972) occur at low concentrations. Carbonate clasts have been replaced by phosphate, glauconite and/or pyrite and it was observed in thin sections that the same authigenic minerals have also precipitated in the intergranular pore space.

The coarse-grained, mineralized bioclastic sand at the basis of the 2.5 m thick Valkenburg Member fines upwards and forms a 0.5 m thick cycle, followed by 2 fining upwards cycles with mineralised sediment at the basis. The upper cycle is 1.5 m thick and has a rather fine-grained, slightly lithified, pure carbonate top with poorly developed flint nodules around spreiten and *Thalassinoides* burrows (Figs. 6.1b, 6.5).

### 6.1.3 Gronsveld Member

The lower part of the Gronsveld Member also consists of fining-upward cycles with a phosphatic-glauconitic-pyritic bioclastic sand at the basis (Fig. 6.5). The mineralised sand of the lowermost cycle is characterised by well developed wavy lamination. Wavy laminated sediment at the basis of the cycles changes upwards via (sub)horizontally laminated sediment towards lithified homogeneously bioturbated, fine-grained, purer carbonate at the top of the cycles.

The upper part of the Gronsveld Member consists of well sorted bioclastic fine sand with low-angle, large-scale wavy lamination (hummocky stratification, cf. Dott & Bourgois, 1982). Flint nodules form laterally restricted curvi-planar layers.

### 6.1.4 Lateral facies changes

The glauconitic cycles of the Valkenburg Member, exposed in quarry ENCI, change laterally towards the South into cycles with flint nodule layers that are very similar to the cycles of the Lanaye Member. At Lanaye (Albert Channel incision) 3 km south of quarry ENCI, flint nodule layers 21, 22, 23 of the Lanaye Member with 23 flint nodule layers (Felder, 1975a,b; Fig. 6.6) are considered to be the chronostratigraphic equivalents of the 3 glauconitic cycles of the Valkenburg Member that cover the Lanaye Member with 20 flint nodule layers at quarry ENCI. Due to the dip of <1° towards the northwest (Francken, 1947), they are at present situated approximately 10 m lower at quarry ENCI than at Lanaye.

## 6.2 Analysis of samples

A 7 m thick succession of grey Chalk (Lanaye Member) and Tuffaceous Chalk (Valkenburg Member), before excavation situated below the groundwater table in the most northwestern part of quarry ENCI and presumably little affected by karst and late diagenetic oxidation, was sampled at dm intervals for element analysis and

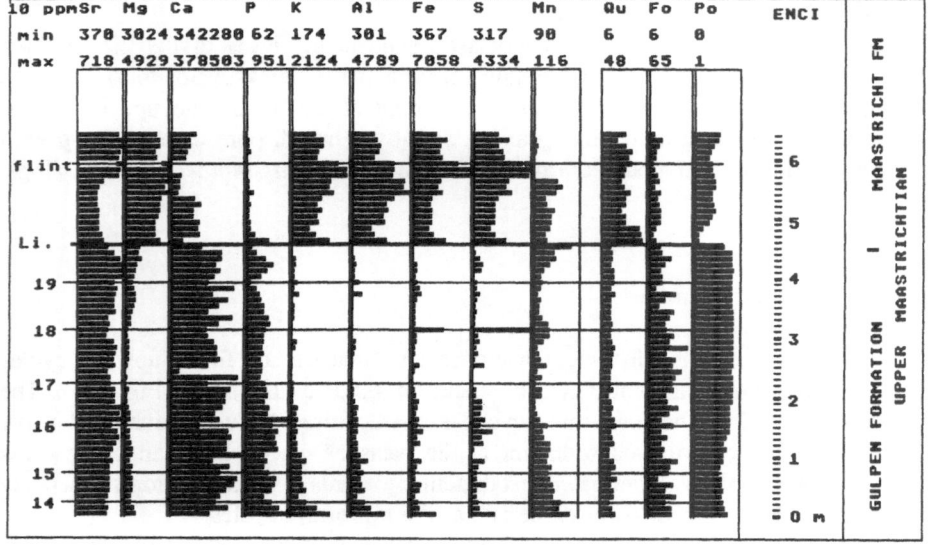

Figure 6.7a - Vertical variation of the element concentrations (ICP). Parts per $10^5$ (except Ca in ppm) in samples taken at dm intervals from the Lanaye and Valkenburg Members (Fig. 6.2). The number of quartz grains (Qu), whole foraminifera (Fo) and estimates of relative porosity (Po) are shown additionally (thin sections) (for data see appendix D).

microfacies study (Fig. 6.2).

Additionally, some sediment for grain-size analysis was sampled at dm intervals between flint layers 12 and 13 of the Lanaye Member (Fig. 4.2) and between 2 flint layers at the top of the Gronsveld Member. In the middle of both intervals a smooth erosion surface is visible, covered by large-scale, wavy laminated, bioclastic silt.

## 6.2.1 Element oxides

The vertical variation of the concentrations of the 9 elements (ICP) (Figs. 6.7a, 6.7b), allows the distinction of 3 groups of elements. The variations of the concentrations of Al, K, Fe and S correlate, so do Ca, Mn, and Sr, while the vertical variation of the Mg and P concentrations correlate neither with the first nor with the second group.

The Al, K and Fe concentrations are a measure of the glauconite concentration and the Fe, S concentrations reflect the concentration of pyrite. The Al, K, Fe and S concentrations correlate negatively with the Ca, Mn and Sr concentrations, that are highest in the purer and most-lithified carbonates and that are a measure of the carbonate (cement) concentration.

| | Sr | Mg | Ca | P | K | Al | Fe | S | Mn | |
|---|---|---|---|---|---|---|---|---|---|---|
| | | -0.701 | 0.898 | 0.802 | -0.882 | -0.887 | -0.882 | -0.884 | 0.853 | Sr |
| | | | -0.044 | -0.626 | 0.818 | 0.838 | 0.753 | 0.751 | -0.497 | Mg |
| | | | | 0.776 | -0.823 | -0.802 | -0.835 | -0.838 | 0.925 | Ca |
| | | | | | -0.926 | -0.921 | -0.937 | -0.937 | 0.715 | P |
| | | | | | | 0.995 | 0.994 | 0.997 | -0.696 | K |
| | | | | | | | 0.981 | 0.986 | -0.669 | Al |
| | | | | | | | | 0.999 | -0.724 | Fe |
| | | | | | | | | | -0.723 | S |
| | | | | | | | | | | Mn |

| Sr | Ca Mn P | Mg K Fe S Al |
|---|---|---|
| Mg | Al K Fe S | Ca Mn P Sr |
| Ca | Mn Sr P | Mg Al K Fe S |
| P | Sr Ca Mn | Mg Al K Fe S |
| K | S Al Fe Mg | Mn Ca Sr P |
| Al | K S Fe Mg | Mn Ca Sr P |
| Fe | S K Al Mg | Mn Ca Sr P |
| S | Fe K Al Mg | Mn Ca Sr P |
| Mn | Ca Sr P | Mg Al K S Fe |

+1     0     -1

Apatite (P,Sr,Ca)
Illite/montmorillonite (Al,Fe,Mg,Mn)
Glauconite (K,Fe,Al) + Pyrite (Fe,S)
Mg,Ca-Carbonate
Sr,Ca-Carbonate
Mn,Ca-Carbonate

Figure 6.7b - Correlation coefficients of the element concentrations and presumed co-occurrence of the various elements in authigenic minerals.

Mg and P correlate poorly with both groups. Mg occurs in glauconite (Harder, 1980; Odin & Matter, 1981) as well as in carbonate (cement). P is present in phosphatic coprolites and vertebrate remains and may occur also as an authigenic precipitate in lithified carbonate and mineralised intraclasts (Berner, 1980).

The phosphatic-pyritic-glauconitic bioclastic sands of the Maastricht Formation are, relative to the Chalk of the Gulpen Formation, characterised by elevated Al, K, Fe and S concentrations. These concentrations show some rhythmic variation in the Chalk of the Gulpen Formation and they decrease upwards. Ca, Mn and Sr concentrations also vary rhythmically and correlate with the occurrence of flint layers, although this correlation is poor.

A clearly negative correlation exists between Mg-concentration and flint concentration. This may reflect early diagenetic lithification of the sediment by means of Mg-carbonate cement precipitation or it may reflect preferential late diagenetic dissolution of Mg-carbonate at the levels of flint nodule growth and consequent precipitation of Mg-carbonates between flint layers.

### 6.2.2 Microfacies

Grainsize variations could hardly be observed in the thin sections of the fine-

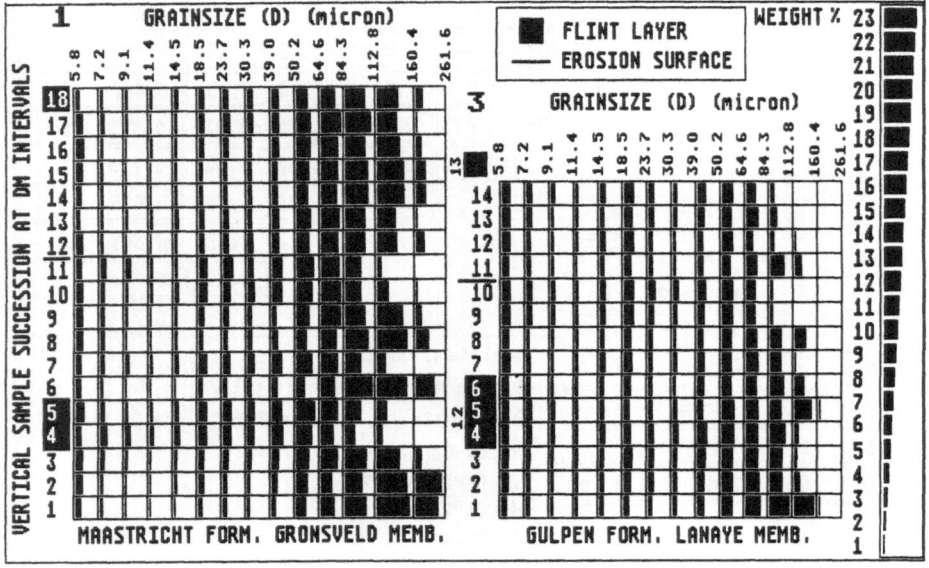

Figure 6.8 - Weight percentages of different grainsize classes (laser particle sizer) in vertical successions of samples from the Lanaye and Gronsveld Members (Fig. 6.2). Note the slightly finer-grained sediment below erosion surfaces and the slightly coarser-grained sediment above the erosion surfaces. Considerable deviations from the fining upwards trend occur in the vicinity of flint nodule layers due to post-depositional late-diagenetic carbonate dissolution and cementation of fine grains into aggregates during flint nodule growth.

grained carbonates. Counts of quartz grains show an elevated concentration in the coarse-grained, mineralised sands of the Valkenburg Member (Fig. 6.7a). In the Chalk of the Lanaye Member a poorly developed rhythmic variation of quartz concentration appears to correlate with Al, K, Fe and S enriched intervals. Counts of whole small benthic foraminifera show elevated concentrations in the intervals characterised by low Al, K, Fe and S concentrations. Porosity is also highest in the Al, K, Fe and S poor intervals.

## 6.2.3 Grainsize

The variation of the grainsize in fine-grained sediment of the Lanaye and Gronsveld Members was measured with a laser particle sizer (Fig. 6.8). Both members contain a distinct 20 micron fraction due to the presence of coccoliths. Due to post-depositional lithification, small grains tend to form larger aggregates, in particular close to flint nodule layers. Nevertheless, somewhat coarser-grained intervals appear to occur above the smooth planar erosion surfaces in the Lanaye and Gronsveld Members, and indicate the presence of fining-upward of grain-size. The presence of a rhythmic variation of the grainsize in the homogeneously bioturbated (Tuffaceous) Chalk is indicated by the detrital quartz concentration and was furthermore proposed on the basis of the rhythmic vertical variation of the concentration of mm-sized bioclasts (Felder, 1986, 1988). Overall, the grainsize increases from fine silt in the Lanaye Member to fine sand in the Gronsveld Member.

## 6.3 Cycle types

The succession at the boundary between the Gulpen and Maastricht Formations exposed in quarry ENCI is characterised by a rhythmic variation of grainsize, structures and authigenic mineral concentration. The character of the cycles varies between two extreme types:

**1** - The cycles of the Lanaye Member (Gulpen Formation) are fine-grained, homogeneously bioturbated coarsening-fining upwards cycles (symmetric cycles, Einsele et al; 1991) of pure carbonates with well developed silica concretion layers and low concentrations of phosphate, glauconite, pyrite and detrital quartz. The sediment has been slightly lithified and hardly compacted. Porosity is high and small benthic foraminifera are complete and well preserved.

**2** - The cycles of the Valkenburg and Gronsveld Members (Maastricht Formation) are coarse-grained, wavy laminated fining-upwards cycles (asymmetric/truncated cycles, Einsele et al; 1991), separated by erosion surfaces. The coarse-grained basis of the cycles is characterised by high concentrations of phosphate, glauconite, pyrite and detrital quartz. The fine-grained top, if preserved, consists of lithified carbonate with poorly developed silica concretions. The coarse-grained mineralised sand at the basis of the cycles has been compacted and the porosity and the concentrations of complete benthic foraminifera are low.

## 6.4 Depositional conditions

The spatial variation of the depositional environment is illustrated by the lateral change of lithofacies of the upper three (21, 22, 23) homogeneously bioturbated fine-grained Chalk cycles with flint nodule layers of the Lanaye Member (Gulpen

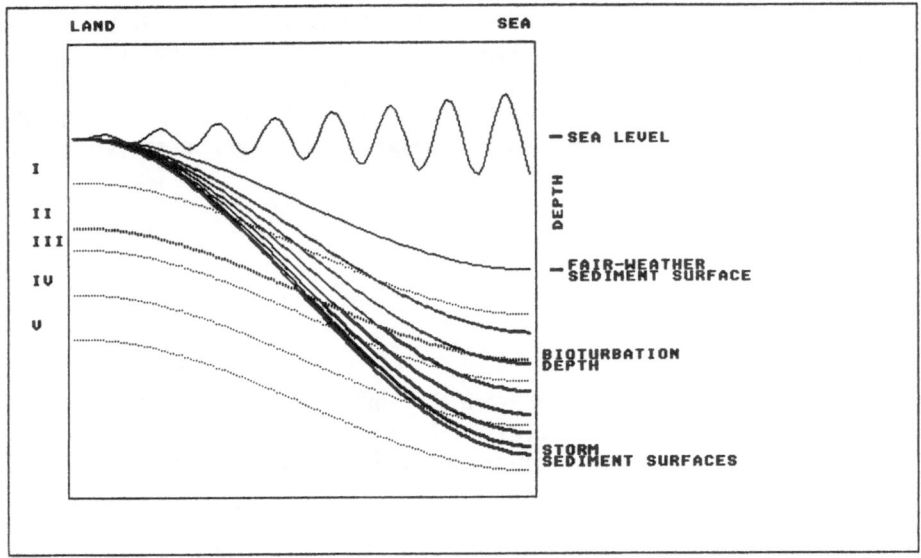

Figure 6.9 - Lateral change of the depositional environment from Lanaye in the South (left) to ENCI 3 km further to the North (right). It is assumed that during the Maastrichtian the seabottom was less than 10 m deeper at ENCI with respect to Lanaye. Hydrodynamic energy (wave-height of sea level) and depth increased towards the North. North dipping fair-weather sediment surface (thin line) with below and parallel, the zone of bioturbative mixing (thick stippled line) and 5 redox zones (thin stippled lines) characterised by manganese-oxide reduction and iron-silicate genesis (II), Fe reduction and iron-sulphide genesis (III), sulphate reduction and Mg-carbonate cement precipitation (IV) and carbondioxide reduction, Mn-carbonate cement and silica authigenesis (V). Storm erosion surfaces (thick lines) dip steeper than the fair-weather sediment surface to the North. Lamination is preserved below the zone of bioturbation at ENCI in the North and otherwise destroyed during bioturbation (thin lines).

Formation) at Lanaye (type 1) to the lower three laminated, rather coarse-grained, glauconitic Tuffaceous Chalk cycles with poorly developed flint nodules of the Valkenburg Member (Maastricht Formation) at quarry ENCI (type 2), 3 km further north and less than 10 m lower at present (Figs. 6.6; 6.9).

If one assumes that the dip of the seabottom during the Late Maastrichtian was less than the present-day dip but also directed towards the NW, then the coarse-grained cycles (type 2) at ENCI have been deposited in higher energetic and deeper water than the fine-grained cycles (type 1) exposed at Lanaye.

During relatively low-energetic fair-weather conditions, fine-grained sediment was bioturbated and mixed in the zone of bioturbation, situated below and parallel to the north-dipping fair-weather sediment surface (Fig. 6.9). However, during storms (Birdsall

& Steward, 1978; Aigner, 1979, 1982; Wright et.al, 1986), part of the sediment was eroded and transported. The thus exposed erosion surface (storm erosion surface) was situated below the fair-weather sediment surface and dipped slightly more than the fair-weather sediment surface towards the North. The depth of storm reworking was least at Lanaye and increased towards ENCI. During waning of storms, the sediment was redeposited in a wavy-laminated (hummocky stratification) fining-upwards sequence, until the fair-weather sediment surface had been restored again. After storms, the zone of bioturbative mixing was re-established and the fining-upwards character and lamination of the storm deposit were destroyed. Only in the more northern and higher energetic environment, where the storm sequence was thicker than the zone of bioturbation, lamination and fining upwards were preserved in the lower part of the fining-upwards sequences formed during waning of storms.

Conceivably, the lateral change of fine-grained homogeneously bioturbated Chalk cycles into coarser-grained, laminated Tuffaceous Chalk cycles with a homogeneously bioturbated top, reflects the lateral increase of the depth of storm reworking and average hydrodynamic energy of the depositional environment.

## 6.5 Early diagenetic conditions

Early diagenesis was characterised by bacterial metabolism, dissolution of detrital minerals and precipitation of authigenic minerals in redox zones that occurred around deep burrows (Bromley et al, 1975; Clayton, 1986; Chapter 3) and below and parallel to the fair-weather sediment surface (Bischoff & Sayles, 1972; Froelich et al., 1979) (Fig. 6.9). These redox zones were:

I - The aerobe shallow zone of bioturbative mixing, carbonate dissolution and manganese oxide precipitation;
II - In the lower part of the zone of bioturbative mixing, the sub-oxic zone of manganese reduction, carbonate dissolution and iron-silicate (smectite-glauconite) precipitation;
III - Below the zone of bioturbative mixing, the sub-oxic zone of iron reduction, carbonate dissolution and iron sulphide precipitation;
IV - Further below, the anoxic zone of sulphate reduction, skeletal opal dissolution and carbonate precipitation; and
V - The deepest anoxic zone of carbondioxide reduction, skeletal opal dissolution and authigenic silica and/or carbonate precipitation.

The erosion of redox zones during a storm, followed by the re-deposition and sorting of authigenic minerals in a fining-upwards sequence according to their fall velocity, changed the distribution of the authigenic minerals, as compared to the distribution below the surface of a seabottom that was not affected by reworking.

The magnitude of the change of the authigenic mineral concentration and distribution as a result of repeated reworking is difficult to determine. For instance, potassium (glauconite) concentration in the coarse-grained bioclastic sands of the Valkenburg Member is up to ten times higher than in the fine-grained bioclastic silts of the Chalk with flint nodules of the Lanaye Member. This does not necessarily indicate a local concentration of coarse-grained glauconitic clasts as the result of transport and sorting during storms, but may also indicate an increased rate of glauconitisation (Burst, 1958; Odin & Matter, 1981; Harder, 1980) as the result of reworking.

For instance, after nascent, instable, fine-grained glauconitic smectite formed at the basis of the zone of bioturbative mixing (redox zone II), it may have been reworked during a storm and re-deposited in the upper part of the fining-upwards sequence, close to the sediment surface. In the oxygenated environment (redox zone I) the instable smectite decomposed and the products (bio)diffused again towards the redox zone below (redox zone II), where they contributed to the further growth of mature, stable, coarse-grained glauconite, that had been re-deposited at the basis of the fining-upwards sequence, situated in the zone of glauconite genesis. The rate of glauconitisation might have been particularly high when the depth of reworking equalled the thickness of the zone of bioturbative mixing. In case reworking was deeper, coarse-grained glauconites, re-deposited at the basis of the fining-upwards sequence, were situated below the zone of glauconitisation after a storm, i.e., in the zone of pyrite genesis (redox zone III) where no further glauconite growth occurred and where glauconite was replaced by pyrite instead.

Another effect of deep storm reworking was the precipitation of carbonate cement and lithification instead of authigenic silica precipitation in the anoxic redox zones of sulphate (IV) and carbondioxide (V) reduction. For instance, the fine-grained homogeneously bioturbated Chalk of the Lanaye Member with well developed flint nodules must have had a high concentrations of fine-grained skeletal opal. The coarser grained homogeneously bioturbated bioclastic silt of the top of the fining-upwards cycles of the Valkenburg Member, that is only poorly silicified, must have had a much lower skeletal opal concentration, hampering the late diagenetic growth of well developed flint nodules (Chapter 4).

The coarser-grained, opal-poor sediment lithified in the redox zones of sulphate and carbondioxide reduction, because the decrease of hydrogen-ion concentration during bacterial metabolism was, at low skeletal opal concentrations, not buffered by $H_4SiO_4$ dissociation, but by the dissociation of $HCO_3^-$, causing the precipitation of carbonate cement. The lithification of the sediment eventually might have led to a decrease of the storm reworking depth, and to a lag between the variation of the storm intensity and the variation of the deposition/erosion rates (Chapter 5).

## 6.6 Cycle genesis

The distribution of grainsize, depositional/bioturbational structures and authigenic mineral concentration in sediment at the bottom of the Chalk Sea was a function of the thickness of the zone of bioturbative mixing, of the thickness of the various redox zones and their reaction rates and of the depth of reworking during storms below the fair-weather sediment surface.

The rhythmic vertical variations of the authigenic mineral concentrations in the (Tuffaceous) Chalk sequences thus indicate a periodic variation of the deposition rates, like the flint nodule layers of the Chalk with Flint (Chapter 5). The rhythmic vertical variation of the grainsize and of the depositional/bioturbational structures witnesses the periodic variation of the hydrodynamic energy and of the depth of storm reworking.

The gradual upward increase of the average grain size (Fig. 6.8) reflects that, besides short-term periodic variations, a longer-term change of the environmental conditions was characterised by a gradual increase of the average hydrodynamic energy.

Concerning the short-term periodic variations, it has been suggested (Chapter 5), that these were the result of periodic variations of climate and in particular of the frequency and intensity of storms. Periodic variations of the Earth's orbital parameters, precession, obliquity and eccentricity, cause periodic variations of the intensity and distribution of insolation, climate and oceanography (Berger, 1979, 1988) and are considered a plausible cause for the rhythmic bedding of (Tuffaceous) Chalk (Hart, 1987; Gale, 1989; Cottle, 1989; Herrington et al., 1991).

Whatever the complex relation between insolation and average storm frequency and intensity might have been during the Maastrichtian, it is assumed that the variation caused the genesis of tempestite cycles during approximately 20 ka precession periods (Chapter 5).

## 6.7 Numerical simulation of the genesis of cycles

In order to increase the understanding of the genesis of the cycles in the (Tuffaceous) Chalk, a numerical model has been developed (Fig. 6.10a,b) that simulates the genesis of rhythmically bedded sequences as a function of periodically varying storm frequency and intensity (see appendix C).

In the model, the position of the fair-weather sediment surface and the grainsize $(G(t))$ of the sediment are defined for a period T during which a sequence of S layers is deposited. It is assumed that the water depth and the sediment grainsize $(G(t))$ are proportional to the hydrodynamic energy $(E(t))$ of the shallow sea. For the sake of simplicity it is assumed that eustatic sea level is constant and that if the average hydrodynamic energy does not vary, then the depth of the shallow sea does not change and the deposition rate (dS/dt) equals the rate of subsidence Sub(t). However, if the hydrodynamic energy varies periodically as a function of the periodic variation of the precession index and the storm intensity, then the depth increases during an increase of the hydrodynamic energy and oceanward sediment transport. Deposition rates thus decrease and eventually erosion occurs when the increase of depth exceeds the subsidence. During a subsequent decrease of the hydrodynamic energy and a consequent decrease of oceanward sediment transport, depth decreases and sediment is deposited at a rate that exceeds the subsidence rate.

$$dS/dt = Sub(t) - C_1*dE/dt$$
$$G(t) = C_2*E(t)$$
$$C_1, C_2 = \text{proportionality factors}$$

While deposition rates and hydrodynamic energy vary periodically, sediment is diagenetically altered in a zone of bioturbative mixing and in various redox zones that occur at, for the sake of simplicity, presumed constant depth and with constant reaction rates, below and parallel to the fair-weather sediment surface. The concentration C(i) of the ith authigenic mineral changes as a function of the reaction rate R(i) and as a function of the time that the sediment resides in the ith redox zone, which is inversely proportional to the deposition or burial rate.

$$dC(i)/dt = R(i)/(1 + ABS(dS/dt))$$

Grainsize and authigenic mineral concentrations in layers at and below the fair-weather sediment surface are redefined after deposition/erosion during each period dt, by averaging (mixing) in the zone of bioturbation, by sorting according to fall velocity in the zone of storm reworking and by reaction in the various redox zones below and parallel to the fair-weather sediment surface.

*Opposite page:*

Figure 6.10a - Results of a numerical model that simulates the grainsize (G), structures (S) and authigenic mineral concentration (CEM) in a cycle deposited during 20 ka. Left box: While sea level and subsidence rate remained constant, the hydrodynamic energy and the depth of the fair-weather sediment surface (thin line) increased-decreased, the bioturbative mixing occurred to a constant depth below the sediment surface (thin line), the sediment was reworked during storms to a maximum depth at 0 ka (thick line), and authigenic mineral precipitation occurred in a redox zone below and parallel to the fair-weather sediment surface (thin stippled lines). Dipping straight lines show the subsidence of sediment deposited at the storm erosion surface at -10, -5, 0, 5 and 10 ka. Right box: Deposited fining-upwards cycle with grainsize (G) and structures (hatched is preserved depositional lamination (see Fig. 6.10b). Blank is homogeneously bioturbated).

Figure 6.10b - As Fig. 6.10a but for higher maximum hydrodynamic energy (0 ka) and depth of reworking. Note that sediment from -10 to 0 ka has not been preserved and that lamination is preserved during periods when depth of storm reworking exceeded the depth of bioturbation (-5 to 5 ka). Authigenic mineral concentration (CEM) is at a maximum in sediment directly below the storm erosion surface of 0 ka, while low in the above reworked sediment.

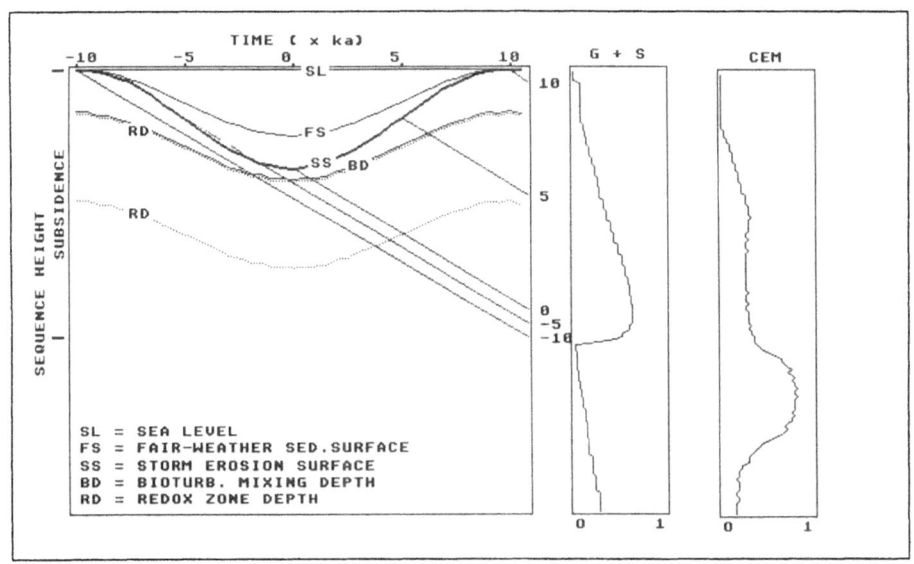

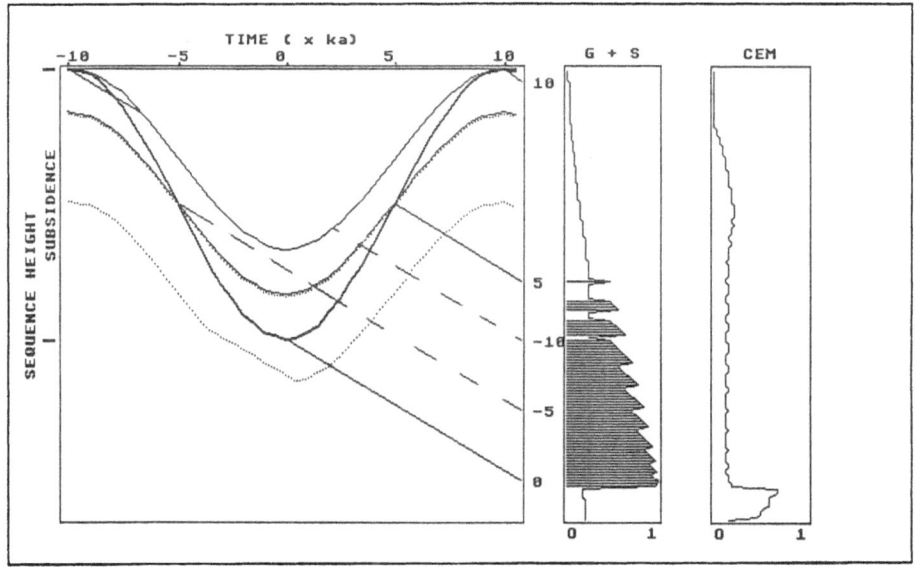

## 6.7.1 Simulation results

With the numerical model (see appendix C) two sequences have been simulated for a 100 ka cycle period with superimposed 20 ka cycles (Figs. 6.11a; 6.11b). For both sequences all parameters are the same, except the maximum depth of the fair-weather (DFS) and storm erosion (DSE) surfaces that reflect the difference between the average hydrodynamic energy as a function of the basin morphology in low energetic (Fig. 6.11a) and high energetic (Fig. 6.11b) environments.

In the model, clay minerals, pyrite, carbonate cement and silica precipitate in respectively, the redox zones II, III, IV and V. All redox zones have a constant thickness, i.e., half the thickness of the zone of bioturbative mixing (BD), except the zone of pyrite precipitation that is only 0.25 BD thick in the model. Clay minerals and pyrite, that tend to form large aggregates or that replace large and heavy bioclasts, are continuously redistributed in a fining-upward sequence with upwards decreasing concentration of these minerals (model sorting factors f1 and f2 set at -1 and -2 respectively, see appendix C).

Carbonate cement concentration is zero in the zone of bioturbative mixing and in the zone of storm reworking. Fine-grained silica concentration is averaged in the zone of bioturbative mixing and increases upwards in the fining-upward zone of storm reworking (f1=1, f2=2). The reaction rates (R) of clay mineral and pyrite precipitation are constant. The reaction rate of carbonate cement precipitation is taken proportional to the grainsize and the reaction rate of silica precipitation is taken inversely proportional to the grainsize, in order to account for the influence of fine-grained skeletal opal concentration.

1 - In the first simulation (low energy, Fig. 6.11a), bioturbation erases depositional lamination and it smooths the vertical variation of grainsize and authigenic mineral concentrations. Authigenic mineral concentrations remain relatively low, due to the constantly relatively high deposition rates. The coarser-grained layers, with highest clay

*Opposite page:*

Figure 6.11a - Result of numerical simulation of the genesis of a rhythmically bedded sequence in a low energy environment. Left box: Constant subsidence rate (straight thin line) during 100 ka. Fluctuation of the deposition rate as a function of the variation of effective hydrodynamic energy and water depth (fair-weather sediment surface, wavy thin line). Below and parallel to the fair-weather sediment surface are the zone of bioturbative mixing with aerobic and suboxic redox zones I, II and the suboxic and anoxic redox zones III, IV and V (stippled wavy lines). Depth of storm reworking never exceeds depth of bioturbative mixing (storm erosion surface, thick wavy line). Sediment layers deposited during moments of maximum hydrodynamic energy at 20 ka intervals (horizontal lines). Right boxes: Grainsize (G) profile with maxima corresponding with highest energy moments, however smoothed by bioturbation. Structures (S) of bioturbation only (blank). Concentrations of authigenic iron-silicates (Cl), iron sulphides (Py), carbonate cement (Ce) and silica (Si) are highest in stratigraphic levels that correspond with the position of the respective redox zones during periods of minimum deposition/erosion rates.

Figure 6.11b - As Fig. 6.11a but with higher effective hydrodynamic energy and storm reworking below the zone of bioturbative mixing. Note the preservation of depositional lamination (S, hatched), the strongly asymmetric character of the cycle succession and the presence of well cemented and poorly silicified "proto-hardgrounds" in the top of coarse-grained fining-upward cycles.

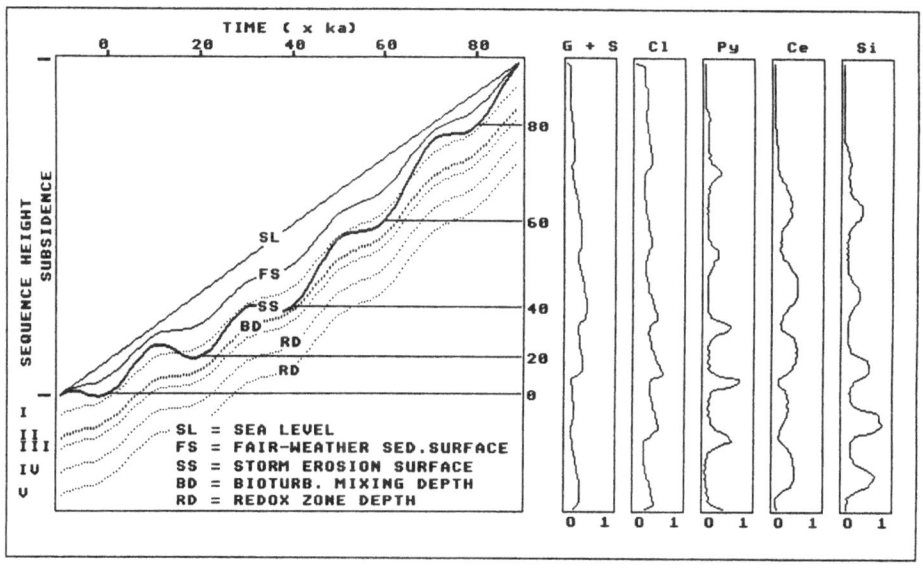

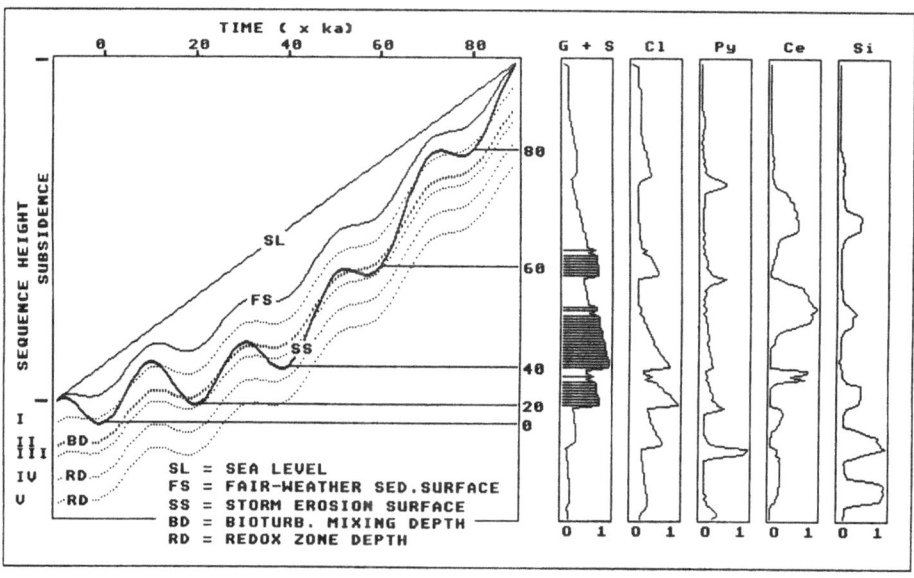

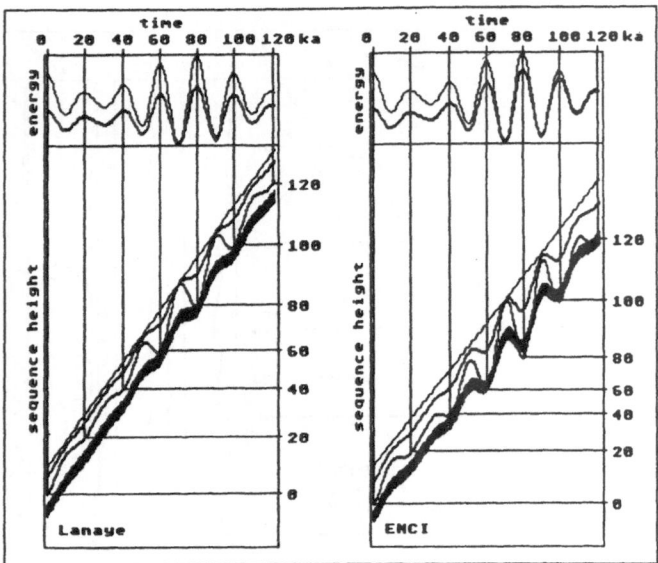

Figure 6.12 - Upper boxes: Periodic variation of the storm intensity (thin lines) and hydrodynamic energy as a function of storm intensity and basin form (thick lines). Lower boxes: Depositional conditions for Lanaye (relatively high deposition rate and low hydrodynamic energy) and ENCI (relatively low deposition rate and high hydrodynamic energy) during a 120 ka period. Thick wavy lines: Fair-weather sediment surface and storm erosion surface below. Wavy hatched zone: Redox zone of silicification (see also Fig. 6.13).

mineral and pyrite concentrations, clearly alternate with the finer-grained layers with a higher concentration of carbonate cement and authigenic silica. The sequence is characterised by a weakly developed cyclic variation of grainsize and authigenic mineral concentration and is comparable with the upper part of the Lanaye Member (Figs. 6.1a, 6.7a).

**2** - In the second simulation (higher energy, Fig. 6.11b), the depth of storm reworking occasionally exceeds the thickness of the zone of bioturbative mixing and well developed fining upwards cycles form. The coarse-grained basis of the cycles is laminated and has a high concentration of clay minerals (smectite/glauconite) and pyrite. The top of

*Opposite page:*

Figure 6.13 - Facies change between Lanaye and ENCI for 120 ka (see Fig. 6.12)
    a: Storm erosion surfaces at 1000 yr intervals, eventually bioturbated (stippled).
    b: Grainsize (dark is coarse-grained, light is fine-grained).
    c: Early diagenetic authigenic silica concentration (dark is high, light is low).

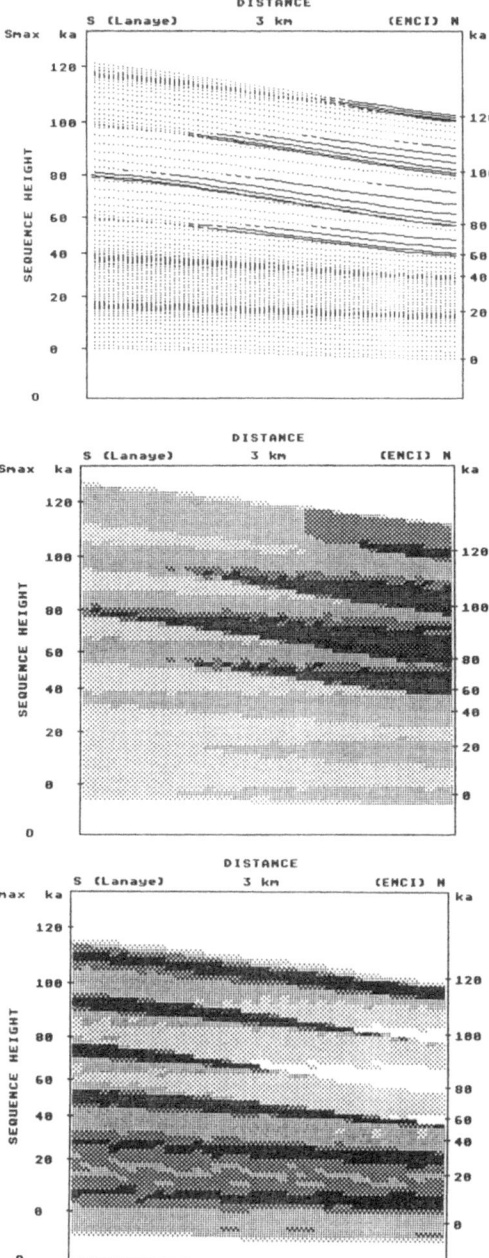

the cycles is fine-grained, homogeneously bioturbated and has a high concentration of carbonate cement, while authigenic silica concentrations are rather low. The sequence is characterised by well developed cyclic variations of grainsize and authigenic mineral concentrations. The strongly asymmetrical cycle is comparable with those in the Valkenburg Member and in the lower part of the Gronsveld Member (Figs. 6.1a, 6.7a).

### 6.7.2  Lateral variation of cycles

The numerical simulation model can also be used to generate a series of sequences, reflecting the gradual lateral change of depositional/early diagenetic conditions. The lateral variation of structures, grainsize/clay mineral concentrations and authigenic silica concentrations has been simulated for a 120 ka period, during which the average hydrodynamic energy increased in time, and in space from south to north, while the subsidence rate was constant in time and space. The results resemble the vertical and lateral change of the lithology along the boundary of the Lanaye Member and Valkenburg Member from Lanaye to quarry ENCI (Figs. 6.6, 6.12, 6.13a, 6.13b, 6.13c).

## Discussion

The numerical model for the genesis of rhythmically bedded (Tuffaceous) Chalk sequences illustrates how cycles with different thickness, structures, grainsize distribution and authigenic mineral concentrations might have formed as a function of the subsidence rate and of the periodic variations of the hydrodynamics and related deposition rates.

The model is based on several assumptions, such as constant sea level, a cyclicity that was caused by periodic variations of the Earth's orbital parameters, a subsidence rate that equalled the average bed thickness divided by the average duration of the precession periods, a deposition rate that varied periodically and that was inversely proportional to the change of hydrodynamic energy, a depth of storm reworking and grainsize that were proportional to the hydrodynamic energy and a constant thickness of the zone of bioturbation and constant thickness and reaction rates of the zones of mineral authigenesis.

These assumptions might be replaced by others without affecting the simulation results of the model. For instance, subsidence can be kept zero while sea-level rises, the thickness of the redox zones might be decreased while reaction rates are increased and a cycle period might be decreased while hydrodynamic energy and depth of storm erosion are increased.

It is doubted that these and other parameters can be defined by means of a more detailed study of the lithology of rhythmic (Tuffaceous) Chalk sequences only. Concerning the definition of the water depth of the Chalk sea, it appears that independently defined parameters have to be taken into account. Opinions concerning water depth are still conflicting. A depth of more than 50 m and deposition below wave base have been proposed mainly because Chalk resembles recent deep sea ooze (Hancock, 1963; Bromley, 1965; Nestler, 1965; Reid, 1968; Håkansson et al, 1974). Others have suggested that Chalk is a "lagoonal" (Binkhorst, 1859; Simien, 1987) or mudflat (Bathurst, 1971, p. 405) deposit. Either the Chalk is a deeper marine sediment, deposited below the wave base, or it is a shallow near-coastal sediment, deposited outside the reach of more open marine energetic waves.

The model may contribute to a solution of the depth problem. For instance, if the presently north dipping coarse-grained cycles at ENCI in the north, were deposited in shallower, higher energetic water than the fine-grained cycles at Lanaye in the south, in line with the deep marine model, then a tectonically induced inversion of the dip has occurred after deposition. The model for the genesis of bedding in Chalk combined with a model for (post) depositional structural evolution thus might resolve the depth problem.

Presently, the model illustrates which possible parameters are at least necessary in order to produce rhythmically bedded (Tuffaceous) Chalk sequences. The definition of the parameters is such that they relate the observed lithologic variation to the dynamics of depositional-early diagenetic conditions. The lithology of (Tuffaceous) Chalk sequences can thus be expressed as the variation of the parameters relative to those of a holotypic sequence, reflecting the dynamics of the Late Cretaceous (Tuffaceous) Chalk sea.

Finally, the model shows that even when the composition of freshly produced marine sediment, a mixture of pore fluid, organic matter, skeletal opal and skeletal silica, does not change, then storm reworking/sorting and early diagenetic bacterial metabolism/mineral authigenesis can cause a considerable deviation from the initial composition of the sediment. Therefore, one should be cautious with conclusions about presumed relations between the variation of the mineralogy/paleontology of sedimentary sequences and the variation of the properties of the seawater during deposition (Buchardt & Jørgenson, 1979; Jørgenson, 1979; Arthur & Dean, 1991).

# Conclusion

Bedding in Chalk sequences reflects a rhythmic variation of grainsize, structures and authigenic mineral concentrations. The rhythmic variation of the lithology is attributed to periodic variations of climate and hydrodynamic conditions that were driven by the orbital influences.

In particular, the periodic variation of the average storm frequency and intensity caused a periodic variation of the hydrodynamic energy and of the depth of the subtropical shallow (Tuffaceous) Chalk Sea. During increasing energy, depth increased and net deposition rates decreased, while during decreasing energy, depth decreased and deposition rates increased.

With the help of a numerical model for the simulation of the genesis of bedding in rhythmical (Tuffaceous) Chalk sequences, different sequences can be generated while varying the subsidence rate, the periodic variation of hydrodynamic energy, the depth of bioturbative mixing, the depth of storm reworking and the depth, thickness and reaction rates of the redox zones of bacterial metabolism below and parallel to the fair-weather sediment surface.

(Tuffaceous) Chalk sequences can not only be described and interpreted according to their resemblance to modern comparable deposits, but with the presented model, the lithology of (Tuffaceous) Chalk sequences can be translated into relatively quantified parameters that describe the dynamics of the depositional-early diagenetic environment and that can eventually be further defined by comparison to independently defined environmental parameters.

# Appendix C - Numerical model

+ = addition  - = substraction  * = multiplication  / = division
m^n = m to the power n
SQR(m) = square root of m
pi=3.14..
sin = goniometric function
A(m,n) = array with m x n sites

GO TO PROCEDURE 1 (in advance defined factors)
GO TO PROCEDURE 2 (depositional/early diagenetic conditions)

PROCEDURE 1 - The subsidence, the depth of the fair-weather sediment surface, the depth of storm reworking, the depth of bioturbation and the depth of a redox zone are defined for a period T, during which S layers are deposited.

        FOR t=0 TO t=T
                E(t)=(0.5+0.5*((A+(1-A)*(0.5+0.5*sin((2*pi/100)*t)))*sin((2*pi/20)*t))
                Fs(t)=(t*(S/T))-DFS*E(t)
                Ss(t)=(t*(S/T))-DSE*E(t)
                Bd(t)=Fs(t)-BD
                Rdmin(t)=Fs(t)-RDmin
                Rdmax(t)=Fs(t)-RDmax
        NEXT t

E(t) = periodically varying hydrodynamic energy (periods 20 and 100 ka, A<1,0<E(t)<1)
Fs(t) = fair-weather sediment surface
Ss(t) = Storm erosion surface
S/T = sequence height divided by simulation time (= constant subsidence rate)
DFS = maximum depth of fair-weather sediment surface
DSE = maximum depth of storm erosion surface (DSE>DFS)
BD(t) = constant depth (BD) of bioturbative mixing below fair-weather sediment surface
Rdmin(t), Rdmax(t) = constant minimum, maximum depth (RDmin, RDmax) of redox zone below fair-weather
            sediment surface
RETURN

PROCEDURE 2 - During the simulation period T, during each step dt sediment is eroded or deposited with a grainsize that is proportional to the hydrodynamic energy (procedure 3). Sediment is mixed in the zone of bioturbive mixing (procedure 4) and eroded, sorted and redeposited to the depth of storm reworking (procedure 5). Authigenic minerals are precipitated in redox zones at a rate that is inversely proportional to the sedimentation rate (procedure 6).

        FOR t=0 TO t=T
                GO TO PROCEDURE 3 (deposition/erosion)
                GO TO PROCEDURE 4 (bioturbative mixing)
                GO TO PROCEDURE 5 (storm reworking)
                GO TO PROCEDURE 6 (authigenic mineral precipitation)
        NEXT t
RETURN

```
PROCEDURE 3 (deposition/erosion)
   FOR s=Fs(t) to s=Fs(t+1)
           IF Fs(t)>Fs(t+1)
                       G(s)=0
                       C(s)=0
           ELSE
                       G(s)=(t*(S/T)-s)/(t*(S/T)-DSE)
           ENDIF
   NEXT s

   G(s) = grainsize of layer s
   C(s) = authigenic mineral concentration of layer s
RETURN
```

```
PROCEDURE 4 (bioturbative mixing)
   FOR s=Fs(t) TO s=Bd(t)
           G_tot=G_tot + G(s)
           C_tot=C_tot + C(s)
   NEXT s
   FOR s=Fs(t) TO s=Bd(t)
           G(s)=G_tot/BD
           C(s)=C_tot/BD
   NEXT s
```

G_tot, C_tot = cumulative grainsize and authigenic mineral concentration in the zone of bioturbative mixing
RETURN

```
PROCEDURE 5 (storm reworking)
   FOR s=Fs(t) TO s=Ss(t)
           G_tot=G_tot + G(s)
           C_tot=C_tot + C(s)
   NEXT s
   FOR s=Fs(t) TO s=Ss(t)
           G(s)=(G_tot/(Fs(t)-Ss(t)))*(1+(f1-f2*((Fs(t)-s)/(Fs(t)-Ss(t)))))
           C(s)=(C_tot/(Fs(t)-Ss(t)))*(1+(f1-f2*((Fs(t)-s)/(Fs(t)-Ss(t)))))
   NEXT s
```

f1,f2 = factors that define upward decrease of grainsize and mineral concentration (coarse-grained authigenics)
         f1=-1,f2=-2) or upward increase of grainsize and mineral concentration (fine-grained authigenics)
         (f1=1, f2=2)
RETURN

```
PROCEDURE 6 (authigenic mineral precipitation)
   FOR s=Rdmin(t) TO s=Rdmax(t)
           C(s)=C(s)+R/(1+ABS(Fs(t+1)-Fs(t)))
   NEXT s
```

   R = reaction rate
RETURN

**Appendix D - Data** of the vertical variation of the element concentration (ppm/10 except Ca in ppm as determined with ICP), the number of whole benthic foraminifera (fo), of quartz grains (qu) and the relative porosity distribution (determined from thin sections) of 67 samples of the Lanaye and Valkenburg Member in the NW part of quarry ENCI as depicted in figure 6.7a.

Valkenburg Member, ENCI NW, dm interval

| x10ppm | Sr | Mg | Ca | P | K | Al | Fe | S | Mn | Qu | Fo | Po |
|---|---|---|---|---|---|---|---|---|---|---|---|---|
| min | 370 | 3024 | 342280 | 62 | 174 | 301 | 367 | 317 | 90 | 6 | 6 | 0 |
| max | 718 | 4929 | 378503 | 951 | 2124 | 4789 | 7058 | 4334 | 116 | 48 | 65 | 1 |
| 67 | 674 | 4364 | 354813 | 91 | 1039 | 2005 | 3189 | 2113 | 91 | 27 | 7 | 34 |
| 66 | 702 | 4509 | 351760 | 132 | 1276 | 2311 | 4353 | 2747 | 91 | 16 | 20 | 35 |
| 65 | 710 | 4543 | 352370 | 124 | 1273 | 2479 | 3573 | 2048 | 93 | 28 | 16 | 45 |
| 64 | 692 | 4355 | 350673 | 159 | 1130 | 2381 | 3466 | 2425 | 91 | 36 | 11 | 52 |
| 63 | 682 | 4375 | 350185 | 93 | 1088 | 2081 | 3791 | 2510 | 91 | 33 | 11 | 54 |
| 62 | 370 | 2695 | 163574 | 63 | 1001 | 3543 | 5423 | 3725 | 44 | 37 | 6 | 49 |
| 61 | 684 | 4929 | 342280 | 105 | 2125 | 3478 | 7058 | 4335 | 90 | 29 | 7 | 49 |
| 60 | 683 | 4915 | 347428 | 106 | 2117 | 3387 | 6813 | 4092 | 90 | 25 | 12 | 44 |
| 59 | 563 | 4804 | 347346 | 141 | 1551 | 4300 | 3639 | 2786 | 105 | 21 | 15 | 43 |
| 58 | 525 | 4703 | 346754 | 144 | 1493 | 4171 | 3331 | 2578 | 107 | 26 | 8 | 43 |
| 57 | 517 | 4825 | 346058 | 234 | 1631 | 4790 | 3745 | 3004 | 103 | 35 | 19 | 49 |
| 56 | 516 | 4608 | 354012 | 130 | 1215 | 3543 | 2626 | 2055 | 97 | 29 | 14 | 48 |
| 55 | 516 | 4507 | 357247 | 156 | 1100 | 2934 | 2435 | 1854 | 98 | 29 | 17 | 46 |
| 54 | 523 | 4512 | 353566 | 127 | 1227 | 3067 | 2558 | 1837 | 102 | 22 | 19 | 43 |
| 53 | 522 | 4437 | 353719 | 149 | 1082 | 2694 | 2553 | 1861 | 101 | 27 | 19 | 47 |
| 52 | 529 | 4312 | 356730 | 180 | 1113 | 2808 | 2587 | 1839 | 101 | 27 | 18 | 55 |
| 51 | 528 | 4312 | 355845 | 168 | 819 | 1847 | 2208 | 1614 | 100 | 46 | 11 | 67 |
| 50 | 547 | 4343 | 356735 | 260 | 1004 | 2005 | 2814 | 1903 | 97 | 48 | 6 | 75 |
| 49 | 557 | 4532 | 348318 | 526 | 1485 | 2954 | 4087 | 2579 | 98 | 51 | 11 | 77 |
| 48 | 625 | 3824 | 365172 | 321 | 416 | 887 | 787 | 433 | 116 | 44 | 17 | 65 |
| 47 | 533 | 4222 | 364983 | 192 | 886 | 2166 | 2211 | 1771 | 94 | 35 | 22 | 46 |

Lanaye Member, ENCI NW, dm interval

| x10ppm | Sr | Mg | Ca | P | K | Al | Fe | S | Mn | Qu | Fo | Po |
|---|---|---|---|---|---|---|---|---|---|---|---|---|
| 46 | 602 | 3677 | 357791 | 199 | 224 | 357 | 398 | 411 | 113 | 7 | 18 | 25 |
| 45 | 649 | 3619 | 367134 | 297 | 217 | 402 | 516 | 428 | 102 | 9 | 22 | 7 |
| 44 | 605 | 3746 | 367223 | 466 | 259 | 424 | 532 | 572 | 104 | 9 | 37 | 7 |
| 43 | 719 | 3576 | 364258 | 616 | 263 | 453 | 690 | 676 | 100 | 8 | 27 | 7 |
| 42 | 700 | 3454 | 363693 | 300 | 232 | 458 | 591 | 610 | 97 | 12 | 18 | 6 |
| 41 | 647 | 3403 | 361024 | 170 | 293 | 605 | 701 | 508 | 94 | 12 | 21 | 7 |
| 40 | 660 | 3417 | 373370 | 292 | 246 | 362 | 807 | 509 | 95 | 7 | 11 | 8 |
| 39 | 651 | 3505 | 362731 | 348 | 326 | 697 | 881 | 607 | 93 | 13 | 44 | 8 |
| 38 | 643 | 4104 | 364357 | 730 | 367 | 369 | 642 | 498 | 94 | 8 | 25 | 7 |
| 37 | 641 | 3337 | 360509 | 438 | 208 | 458 | 441 | 501 | 93 | 7 | 17 | 7 |
| 36 | 640 | 3419 | 369723 | 367 | 208 | 369 | 604 | 440 | 95 | 8 | 35 | 6 |
| 35 | 596 | 3319 | 368206 | 518 | 253 | 384 | 605 | 447 | 95 | 8 | 26 | 6 |
| 34 | 608 | 3434 | 366102 | 546 | 206 | 345 | 545 | 440 | 95 | 8 | 30 | 6 |
| 33 | 599 | 3281 | 364983 | 349 | 179 | 312 | 2196 | 2222 | 100 | 8 | 41 | 5 |
| 32 | 671 | 3421 | 369095 | 467 | 175 | 311 | 683 | 561 | 100 | 8 | 25 | 5 |
| 31 | 673 | 3373 | 365259 | 485 | 208 | 505 | 571 | 516 | 96 | 13 | 50 | 5 |
| 30 | 671 | 3360 | 364118 | 213 | 213 | 563 | 574 | 542 | 93 | 12 | 45 | 6 |
| 29 | 639 | 3230 | 366294 | 371 | 208 | 419 | 596 | 433 | 94 | 9 | 40 | 5 |
| 28 | 596 | 3173 | 366765 | 448 | 208 | 348 | 848 | 754 | 98 | 8 | 56 | 5 |
| 27 | 554 | 3107 | 360780 | 493 | 233 | 387 | 565 | 417 | 95 | 8 | 44 | 6 |
| 26 | 554 | 3222 | 371146 | 496 | 243 | 363 | 525 | 414 | 95 | 11 | 43 | 6 |
| 25 | 559 | 3176 | 369123 | 515 | 251 | 424 | 587 | 479 | 96 | 6 | 43 | 6 |
| 24 | 557 | 3189 | 369807 | 451 | 244 | 441 | 640 | 472 | 92 | 10 | 38 | 6 |
| 23 | 597 | 3235 | 365444 | 508 | 227 | 469 | 816 | 691 | 92 | 11 | 50 | 6 |
| 22 | 636 | 3308 | 366657 | 527 | 231 | 370 | 742 | 676 | 92 | 9 | 34 | 10 |
| 21 | 681 | 3337 | 367614 | 293 | 314 | 607 | 812 | 627 | 94 | 15 | 41 | 11 |
| 20 | 687 | 3289 | 372598 | 714 | 291 | 607 | 763 | 681 | 96 | 10 | 21 | 12 |
| 19 | 664 | 3389 | 370326 | 688 | 327 | 582 | 1098 | 1026 | 97 | 11 | 31 | 12 |
| 18 | 640 | 3407 | 378591 | 468 | 303 | 577 | 877 | 740 | 101 | 14 | 27 | 17 |
| 17 | 624 | 3439 | 367745 | 387 | 394 | 630 | 577 | 457 | 101 | 8 | 24 | 18 |
| 16 | 620 | 3404 | 367348 | 543 | 394 | 735 | 582 | 404 | 100 | 14 | 36 | 31 |
| 15 | 617 | 3416 | 374745 | 710 | 386 | 824 | 719 | 600 | 102 | 9 | 34 | 28 |
| 14 | 617 | 3429 | 373715 | 670 | 415 | 745 | 635 | 532 | 101 | 18 | 15 | 28 |
| 13 | 593 | 3351 | 365260 | 692 | 362 | 766 | 1020 | 838 | 103 | 9 | 42 | 25 |
| 12 | 613 | 3374 | 365816 | 593 | 375 | 897 | 873 | 518 | 104 | 14 | 28 | 25 |
| 11 | 600 | 3471 | 370650 | 755 | 655 | 871 | 795 | 570 | 107 | 8 | 28 | 21 |
| 10 | 594 | 3537 | 362209 | 914 | 881 | 1642 | 1799 | 1133 | 113 | 15 | 40 | 15 |

# 7 The Genesis of Hardgrounds

### Abstract

Maastrichtian planar-parallel bedded Chalk and wavy bedded bioclastic carbonate sandstone (Tuffaceous Chalk) of the Gironde Estuary (SW France) and of the Maastricht area (SE Netherlands) are characterised by flint nodule layers or by layers that have been lithified by carbonate cement. Silica and/or carbonate cement precipitated during early diagenesis as a result of bacterial metabolism in anoxic redox zones that were situated several dm below and parallel to the sediment surface. Bedding was a function of periodically varying deposition rates and depth of storm reworking that followed periodic variations in the Earth's orbital parameters, climate and oceanography. Well developed lithification by carbonate cement occurred during erosion and low deposition rates in a relatively high energetic depositional environment. When the cemented layers were eroded and covered again during waxing and waning of storms, wavy erosion surfaces formed that gradually changed form and that slowly migrated during periods of several ka. When lithified layers were continuously exposed, then bored, encrusted and mineralised hardgrounds formed. The genesis of wavy bedded silicified or lithified cycles in (Tuffaceous) Chalk is discussed and simulated with the help of a numerical model.

## Introduction

Several tens of metres thick sequences of Maastrichtian (Dumont, 1849; Dordonian s.s. Coquand, 1857) subtropical, shallow marine bioclastic carbonates are exposed in quarries (120 m, Felder, 1975a,b) near Maastricht (SE Netherlands) and in a cliff-section (45 m, Séronie-Vivien, 1972) along the Gironde Estuary (SW France) (Fig. 2.1).

The successions are characterised by a gradual coarsening upward from coccolithic mudstones (Chalk) towards coccolithic bioclastic sandstones (Tuffaceous Chalk). The successions consist of laterally continuous, planar-parallel beds, thickening upwards from a mean of a dm to more than 2 metres.

Figure 7.1 - Approximately 20 m of Campanian/Maastrichtian (Séronie-Vivien, 1972) pelloidal coccolithic mudstones exposed at Le Caillaud (Gironde, SW France). Laterally continuous planar-parallel beds thicken upwards. Thin beds at the basis consist of couplets of relatively coarse-grained, fossiliferous glauconitic carbonate and fine-grained, lithified pure carbonate. The beds at the top consist of couplets of thin, poorly lithified carbonate and of thick, lithified pure carbonate with poorly developed diffuse flint nodules.

The thin-bedded, fine-grained Chalk at the basis has been homogeneously bioturbated and the bedding is defined by a rhythmic variation of the grainsize (Felder, 1986, 1988) and by a rhythmic variation of the concentration of authigenic minerals (e.g. smectite-glauconite, pyrite, carbonate cement and authigenic silica; Bromley et al., 1975; Clayton, 1986; Zijlstra, 1987, 1989).

The thicker beds of the coarse-grained Tuffaceous Chalk are fining upwards cycles,

Figure 7.2 - Approximately 20 m of Maastrichtian Chalk with Flint of the Gulpen Formation (Lixhe 3 and Lanaye Members; Felder, 1975a,b) with the Nivelle Horizon (arrow). The sequence coarsens upwards from coccolithic mudstones into coccolithic bioclastic siltstones and is exposed in quarry North, 10 km south of Maastricht (SE Netherlands). Laterally continuous planar-parallel beds with layers of silica concretions (flint) thicken upwards.

bounded by wavy erosion surfaces. The coarse-grained basis of the cycles consists of (cross)laminated phosphatic-glauconitic-pyritic bioclastic sand, while the fine-grained top consists of homogeneously bioturbated, lithified and/or silicified bioclastic silt. The top-surface of the Tuffaceous Chalk cycles may be a bored, encrusted and/or mineralized and it formed a rocky seabottom (hardground) during times of non-deposition (Voigt, 1929, 1959, 1974).

The spatial-temporal succession of deposition, bioturbation, lithification, boring, encrusting and mineralisation of the sediment that forms a hardground has received considerable attention, because it informs about the environmental conditions during periods of non-deposition (Bromley, 1968; Purser, 1969; Bathurst, 1971; Fursich, 1979; Wilson, 1990).

The question, why some beds are poorly lithified and others are well lithified hardgrounds, has received far less attention. Several examples of lithofacies characterised by lithification from the Maastrichtian (Tuffaceous) Chalk of Maastricht and the Gironde Estuary will be discussed and a model is presented that relates the environmental parameters (e.g. tectonics, climate and oceanography) to deposition rates, sediment grain size and degree of lithification.

# 7.1 Litho-/biofacies

### 7.1.1 Lower part of the successions

The fine-grained Chalk and bioclastic silt at the basis of the Maastrichtian sequences has been homogeneously bioturbated and is planar-parallel bedded. At Maastricht, the bedding is defined by layers of cryptocrystalline quartz concretions (flint, Buurman & van der Plas, 1971), while along the Gironde estuary, the bedding is defined by more or less lithified layers (carbonate cement) with poorly developed silicification (Figs. 7.1, 7.2). The sediment coarsens and beds thicken upwards. The thinner beds at the basis of the successions are characterised by relatively high concentrations of smectite/glauconite.

### 7.1.2 Middle part of the successions

The bioclastic silt and fine-grained sand in the middle part of both sequences is partly bioturbated and partly cross-bedded. The cycles of the Gironde sequence are fining-upward cycles, bounded by erosion surfaces (Figs. 7.3, 7.4). The lower part of the cycles is characterised by wavy (hummocky cross-bedding; Dott & Bourgois, 1982) and by curvi-planar erosion surfaces, that closely succeed each other laterally in a preferred direction. The top of the cycles has been homogeneously bioturbated and lithified.

The cycles of the middle part of the Maastricht succession (Figs. 7.5, 7.6) are fining-upwards cycles, bounded by wavy erosion surfaces. Hummocky cross-lamination has only been preserved locally in the troughs of the wavy erosion surfaces. The sediment of the upper part of the cycles has been homogeneously bioturbated. Lithification is poorly developed. Instead oblique flint nodule layers occur that have a lateral extension of half the wave-length of the wavy erosion surfaces and a dip that is slightly less than the dip of the erosion surfaces.

*Opposite page:*

Figure 7.3 - (top) Maastrichtian pelloidal coccolithic mudstones (6 m) with 6 cycles bounded by wavy erosion surfaces (arrows), exposed near Talmont (Gironde Estuary). The beds at the basis are relatively thin, well lithified and strongly eroded. Beds at the top are thicker, poorly lithified and hardly eroded. The beds consist of a hummocky or trough cross-bedded basis and a homogeneously bioturbated top.

Figure 7.4 - (bottom) Two metres thick bed in the middle of the Talmont sequence (Fig. 7.3). Trough cross-bedded basis, covering eroded, bioturbated and lithified layer, is covered by homogeneously bioturbated and lithified top.

126

## 7.1.3  Upper part of the successions

The bioclastic sand of the Tuffaceous Chalk in the upper part of both sequences (Figs. 7.7, 7.8) is characterised by well developed fining-upwards cycles. The thinner beds are homogeneously bioturbated, while the thicker beds show cross-bedding and (cross)lamination in the coarse-grained lower part (Figs. 7.9, 7.10).

Curvi-planar erosion surfaces succeed each other laterally in a preferred direction. The mean grainsize decreases upwards in a cycle, while the lateral distance between erosion surfaces and the steepness of the erosion surfaces increases upwards in a cycle. The top of the cycles has been homogeneously bioturbated, lithified and partly eroded. The top of thick cycles is poorly lithified and nodular lithification around *Thalassinoides* burrows typically occurs at some depth below the erosion surface (Fig. 7.10). Laterally, the cycles may thin and the erosion surface may descend, while the nodular lithification changes into massive lithification. The morphology of the erosion surface changes in that case from smooth, via capricious (enlarged *Thalassinoides* burrows), encrusted, bored and mineralised towards smooth with eroded boreholes.

The erosion surface is covered by coarse-grained glauconitic and ferruginous bioclastic sand (Fig. 7.10), which forms the basis of the next cycle. Bioclasts are the skeletal remains of bryozoa, red algae, echinoderms and large benthic foraminifera (Fig. 7.11). In the depressions of the erosion surface, above the well lithified cycle top, sediment is very coarse-grained and contains exhumed (Voigt, 1968) lithified tubular nodules formed around burrows of *Thalassinoides*, large encrusted, bored and mineralised pebbles of reworked lithified sediment (Fig. 7.12) and the skeletal remains or molds (dissolved aragonite) of epi-/endolithic rudists, red algae and hermatypic colonial corals.

*Opposite page:*

Figure 7.5 - Three metres thick succession of fine-grained bioclastic sand of the boundary between the Schiepersberg and Emael Member (Romontbos Horizon, arrows) exposed in quarry ENCI (Maastricht). An approximately 10 m wide and 1 m deep trough of the wavy erosion surface of the Romontbos Horizon is filled with large-scale, hummocky trough cross-laminated sediment. At least 4 platy flint nodule layers dip towards the right (west), i.e., oppositely to the dip of the Romontbos Horizon. They have been eroded at the right-hand side and are succeeded by barely visible, slightly steeper right-dipping erosion surfaces, which amalgamate with the left-dipping Romontbos Horizon below.

Figure 7.6 - Flint nodule distribution in a 6 m high, 155 m long wall of quarry NEKAMI ('t Rooth, the Netherlands). High energetic basis of the Schiepersberg Member (Maastricht Formation) with 5-6 flint nodule layers is characterised by hummocky cross-stratification, curvi-planar erosion surfaces and flint nodule layers.

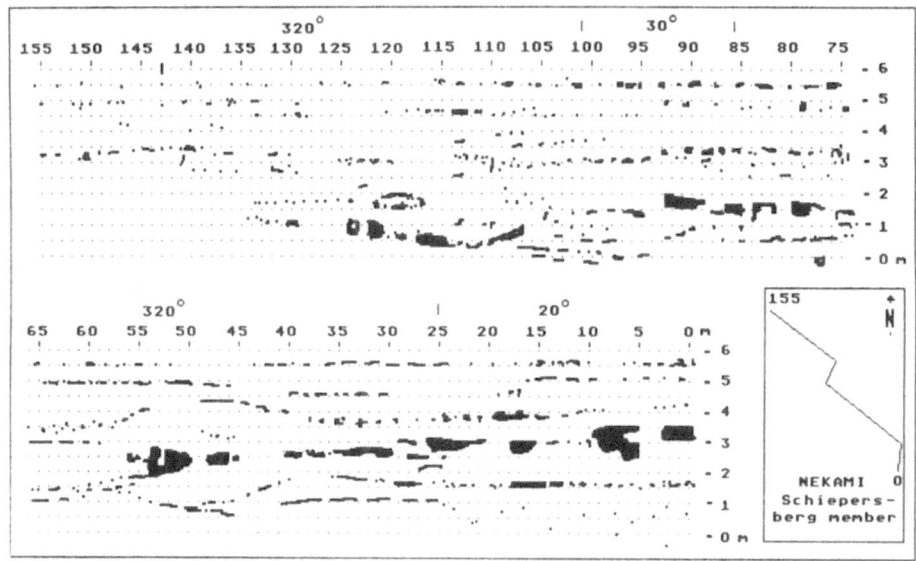

Figure 7.7 - Approximately 25 m of Maastrichtian bioclastic sandstone with bryozoa and rudists exposed at Meschers-sur-Gironde (Plage des Nonnes). Upper beds with wavy erosion surfaces.

Figure 7.8 - Approximately 25 m of Maastrichtian bioclastic sandstone (Emael, Nekum and Meerssen Members bounded by Laumont and Caster Horizons (arrows)), with bryozoa and rudists exposed in quarry ENCI (Maastricht). Upper beds with wavy erosion surfaces.

Figure 7.9 - 3 m thick succession of bioclastic sand near the basis of the Meerssen Member. The ceiling of the cave is formed by a nodular lithified layer (Fig. 7.10) that occurs in the fine-grained top of a cycle. The cycle has been partly eroded and the erosion surface (large arrow) is covered by a fining upwards sequence with a hummocky/trough cross-bedded basis and a top that is again homogeneously bioturbated, lithified and eroded (small arrow).

## 7.2 Syn-depositional lithification of Chalk

The Gironde and Maastricht sequences are very similar with regard to thickness, the upward coarsening of grainsize, the upward thickening of beds and the fossil content. The difference lies in the fact that the fine-grained, thin-bedded lower part of the Maastricht sequence is silicified, while the fine-grained, thin-bedded lower part of the Gironde sequence is lithified by carbonate cement.

The carbonate cement consists of blocky microcrystalline low-magnesium calcite. However, the post-depositional karstification, dissolution and reprecipitation of the porous carbonates hampers the recognition of the syn-depositional cement. Nevertheless, given the hardgrounds in the upper part of the sequence, the lithification of the lower part of the Gironde sequence is considered also to have occurred during deposition.

It has been suggested (Chapter 3, 4) that carbonate cement and/or authigenic silica

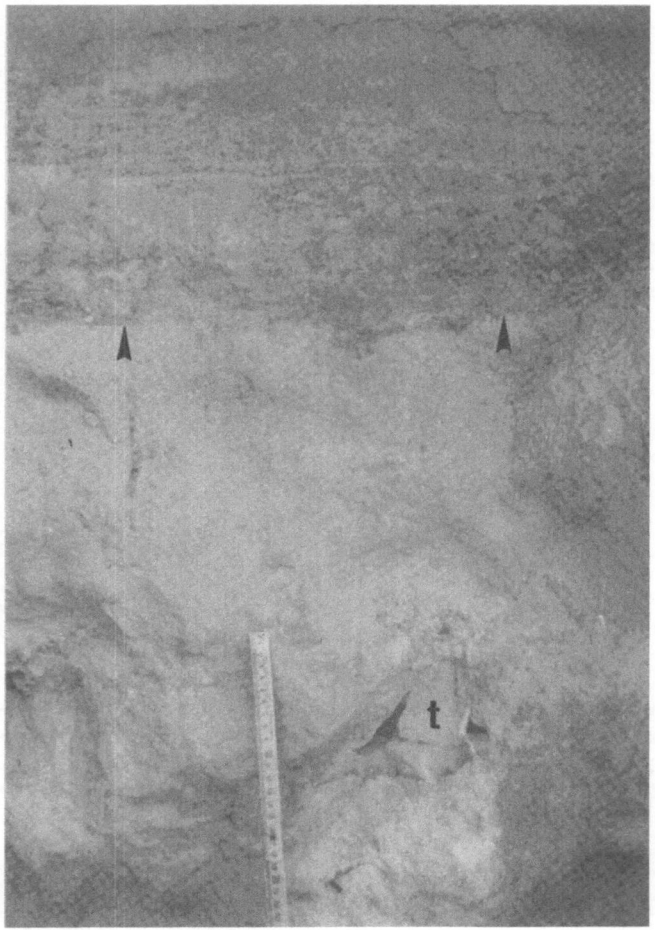

Figure 7.10 - Detail of the basis of Fig. 7.9. Homogeneously bioturbated fine-grained top of a cycle with nodular lithification around deformed *Thalassinoides* burrows (t). Part of the cycles has been eroded and the erosion surface (arrow) was covered by laminated, coarse-grained, glauconitic, pyritic, bioclastic sand.

precipitated in the anoxic redox zones of sulphate and/or carbondioxide reduction as a result of bacterial metabolism. This hypothesis appears to be supported by the isotope record of carbonate concretions (Hudson, 1977; Gautier & Claypool, 1984; Raiswell, 1987). The genesis of $H_2S$ and $CH_4$ required hydrogen ions and this led to the dissociation of $H_4SiO_4$, the polymerization of $H_3SiO_4^-$ and the precipitation of $SiO_2$ in the Chalk of Maastricht with a high skeletal opal concentration. While in the Chalk of the Gironde

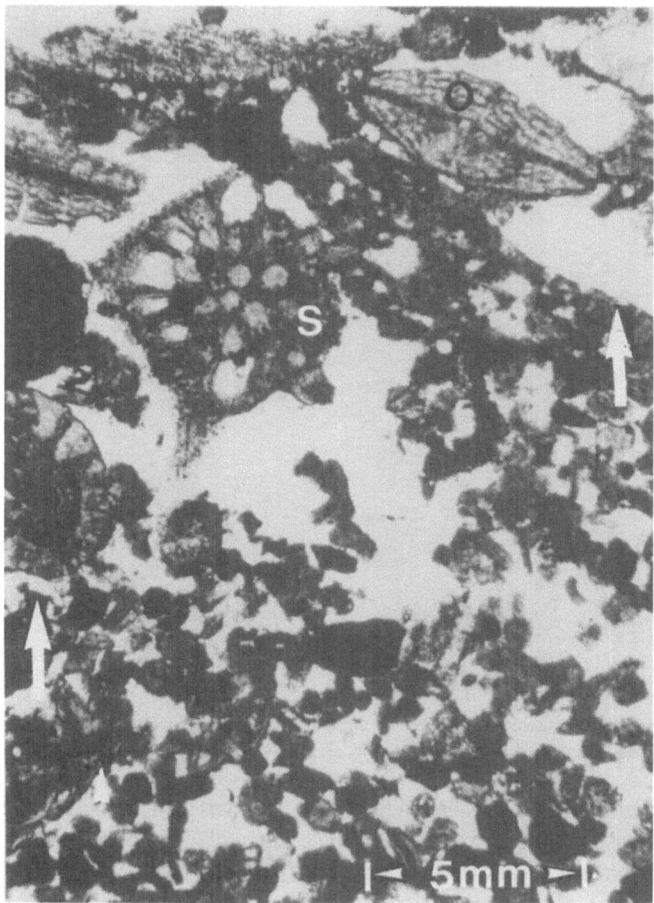

Figure 7.11 - Thin section (6x) of very porous, coarse-grained bioclastic sand of the lithified top of the
Nekum Member and coarse-grained basis of the Meerssen Member (Caster Horizon, arrows). The coarse-
grained bioclastic sand of the basis of the first cycle of the Meerssen Member contains large benthic
foraminifera (Orbotoides and Siderolites), bryozoa and echinoid fragments. Grains are moreover filled
and/or replaced by iron-oxide (presumably oxidized pyrite).

sequence with a low skeletal opal concentration, the genesis of $H_2S$ and $CH_4$ led to the
dissociation of $HCO_3$, the precipitation of $CaCO_3$ and to syn-depositional lithification
of the Chalk. Likewise, the upward increase of grainsize and the decrease of fine-grained
skeletal opal concentration in the Maastricht sequence is reflected by the decrease of
the flint concentration and the appearance of layers that have been lithified by carbonate
cement (hardgrounds) instead of silicified.

The concentration of authigenic silica and/or carbonate cement was not only a function

Figure 7.12 - Large (3 dm) reworked boulder of lithified bioclastic sand of the Meerssen Member, that has been bored and encrusted by algae, bryozoa, serpulids and brachiopods at all sides.

of the bacterial metabolism and the reaction rates, but was also a function of the residence time of the sediment in the anoxic redox zones, thus inversely proportional to the burial or the deposition rates (Berner, 1980; Chapter 5, 6). The rhythmic variation of flint and/or carbonate cement concentration in the planar-parallel bedded (Tuffaceous) Chalk sequences was interpreted to reflect a periodic variation of the deposition rates, caused by the Earth's orbital parameters (precession 20 ka, eccentricity 100 ka), climate and oceanography (Berger, 1979; Hart, 1987; Cottle, 1989; Herrington et al., 1991; Chapter 5). And it has been suggested that in particular the periodic variation of the storm frequency and intensity has caused the bedding and the vertical rhythmic variation of grainsize, structures and authigenic mineral concentrations in (Tuffaceous) Chalk (Chapter 6).

Assuming constant sea level, the grainsize ($G(t)$) and the rate of sediment deposition ($dS/dt$) were a function of the subsidence rate ($Sub(t)$) and of the periodically varying hydrodynamic energy ($E(t)$).

$$dS/dt = Sub(t) - C_1 * dE/dt$$
$$G(t) = C_2 * E(t)$$
$$C_1, C_2 = \text{proportionality factors}$$

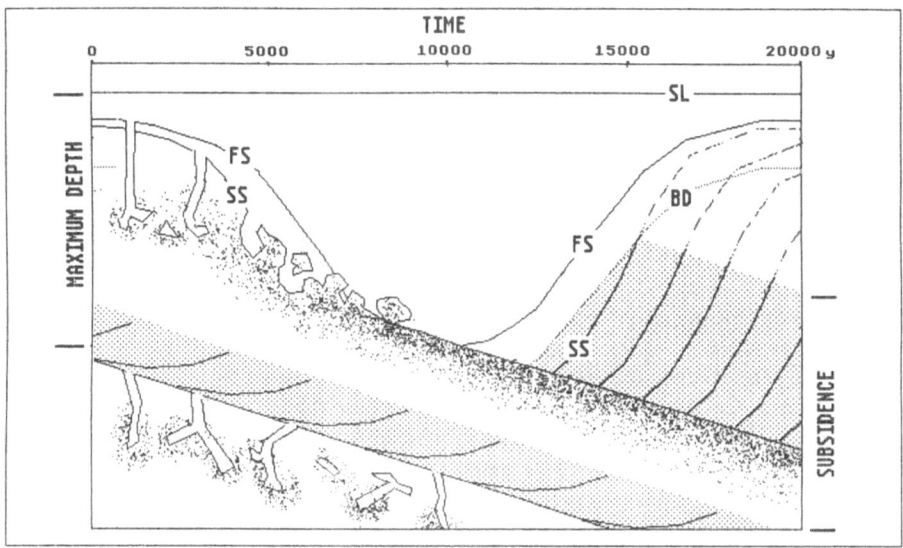

Figure 7.13 - The succession of depositional environments during a 20 ka precession period, characterised
by constant sea level (SL), constant subsidence rate (oblique time lines) and an increase-decrease of
hydrodynamic energy and depth of the fair-weather sediment surface (FS), following a periodic variation
of climate and storm frequency-intensity. During 0-5 ka, depth increase is proportional to subsidence,
deposition rates are zero and sediment consolidates, is bioturbated by crustaceans and lithifies in the
anoxic redox zones. During 5-10 ka, depth increase exceeds subsidence and sediment is eroded. The
fair-weather sediment surface coincides with the storm erosion surface (SS) and while the deep sediment
is eroded and further lithifies, a continuously exposed hardground appears. During 10-20 ka, hydrodynamic
energy, water depth and depth of storm reworking decrease again. A fining-upward sequence is deposited
of which the lower part is laminated (grey shaded) and the upper part is mixed by bioturbation (BD).
Also depicted is a relatively low energetic cycle deposited during the previous precession period.

The rate (R(s)) of authigenic mineral precipitation is assumed to have been constant
at a fixed depth (S(t)-s) below the sediment surface (S(t)). The change of the concentration
of authigenic minerals (dC(s)/dt) in a layer of sediment (s) situated in the relative part
of the redox zone is assumed to have been inversely proportional to the deposition rate
and proportional to the reaction rate (R(s)).

$$dC(s)/dt = R(s)/(1 + ABS(dS/dt))$$

According to the model, well lithified layers formed during the increase of the average
storm intensity and related increase of hydrodynamic energy, when the increase of water
depth equalled the subsidence rate and when the deposition rate was consequently zero.
During a subsequent further increase of the hydrodynamic energy, the sediment that

had been lithified at some depth below the fair-weather sediment surface, was repeatedly eroded, exhumed and exposed during storms, and again covered by a fining-upwards storm layer during waning of the storms. When the hydrodynamic energy reached its maximum during the precession period, then the lithified layer was continuously exposed and developed into a bored and encrusted hardground. During the subsequent decrease of the intensity of storms and average hydrodynamic energy, the hardground was covered by a fining upwards sequence again and a new cycle was formed during the next precession period (Fig. 7.13).

The genesis of only slightly lithified, fine-grained and siliceous, parallel-bedded (Tuffaceous) Chalk sequences with a rhythmic variation of grainsize, structures and authigenic mineral concentrations, could be simulated with a numerical model (Chapters 5, 6).

The simulation of the genesis of well lithified, coarser-grained (Tuffaceous) Chalk sequences is more complex (Fig. 7.14). When the hydrodynamic energy is high and the erosion rate exceeds the subsidence rate, then the erosion rate is not only a function of the storm intensity and hydrodynamic energy but also of the decreased erodability of the previously lithified sediment. The lithification prevents deep erosion during high-energy periods and is expected to suppress the cycle thickness variation in sequences formed during a succession of precession periods with different storm intensity maxima (Chapter 5).

## 7.3  Relation between lithification and hydrodynamics

In planar, parallel-bedded (Tuffaceous) Chalk sequences, the bed thickness and vertical variation of the grainsize, structures and authigenic mineral concentrations are suggested to have been a function of the temporal variation of the storm intensity and hydrodynamic energy. In the cross-bedded Tuffaceous Chalk sequences a superimposed spatial variation of the hydrodynamic energy, deposition rates and early diagenetic conditions must have influenced the genesis of bedding.

The lateral succession of wavy and curvi-planar erosion surfaces at the coarse-grained basis of the Tuffaceous Chalk cycles reflects the lateral migration of wavy storm erosion surfaces during several ka long periods of maximum hydrodynamic energy characterised by successions of relatively strong storms.

In the middle part of the Gironde sequence (Talmont, Figs. 7.3, 7.4), wavy storm erosion surfaces, formed at presumably more than 100 y intervals, 'inherit' the morphology of the precursor wavy storm erosion surfaces. The oblique-dipping flint layers in the middle part of the Maastricht Formation (quarry ENCI and Nekami, Figs. 7.5, 7.6), which also formed during periods of relatively low deposition rates in the anoxic redox zone below and parallel to the fair-weather sediment surface, indicate that the fair-weather sediment surface and the redox zones below were also wavy and that also these more or less retained their form over periods of presumably several 100 to 1000 years, while

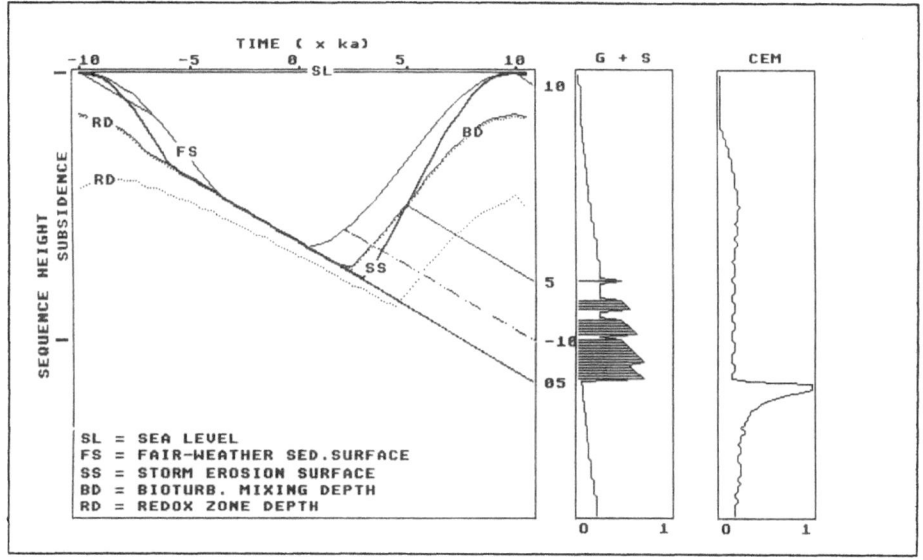

Figure 7.14 - The same as in Figure 6.10b, but with lithification. The increase of the depth of the fair-weather sediment surface, the depth of bioturbation, and the increase of the depth of storm reworking and the thickness of the redox zone have been diminished proportionally to the degree of lithification. Note that the depth of reworking is less than in the case without lithification (Fig. 6.10b) and that a well lithified hardground has formed. Depositional structures deposited between 0 and 5 ka have escaped bioturbative mixing and are preserved. Grainsize reflects hydrodynamic conditions during deposition. Note that much coarser-grained sediment is found at the basis of natural cycles, where coarse-grained sediment, a relict of pre 0 ka reworking, is preserved, while it has been eroded in the simulation model.

migrating laterally.

The preservation of the morphology of the fair-weather sediment surface, zones of bioturbation and authigenesis and of the storm erosion surface, suggests that bed morphology was not only a function of the instantaneous hydrodynamic conditions during storms, but also, to a large extent, a function of the already present bed morphology, defined during previous storms. There appears to be a "memory" for bed form, and it is suggested that the increase of preservation potential of bed morphology was due to syn-depositional lithification.

## 7.4 Numerical simulation of wavy bedding

The environmental parameters that defined the genesis of parallel bedded cycles also

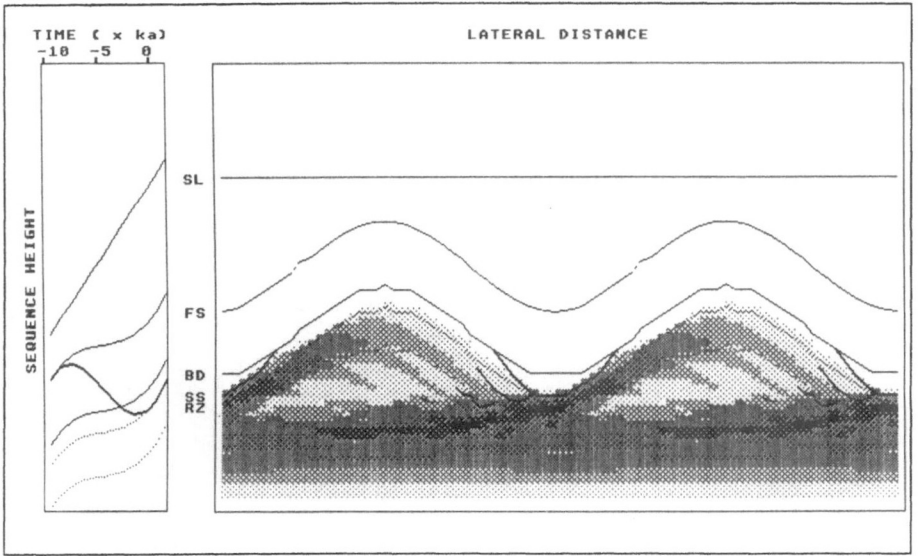

Figure 7.15 - Result of the simulation of the genesis of the wavy bedded middle part of the Maastricht sequence (Fig. 7.5). Left box: The position of the sealevel (thin straight line, SL), fair-weather sediment surface (thin wavy line, FS), the depth of the zone of bioturbation (thin wavy line, BD), the depth of the zone of authigenic silica precipitation (stippled wavy lines, RZ) and the depth of the storm sediment surface (fat wavy line, SS) during a 10 ka period, from low to high hydrodynamic energy conditions. Right box: sea level, wavy fair-weather sediment surface, zone of bioturbative mixing, redox zone and partly bioturbated storm erosion surface during moment of highest hydrodynamic energy. During the preceding 10 ka period, hydrodynamic energy and wave-heights increased, while the lateral migration rates of the waves decreased. Simultaneously sediment was silicified at constant rate in the redox zone below and parallel to the fair-weather sediment surface (dark=high concentration, light=low concentration). Note the left-(east-) dipping proto-flint layers below the left dipping "Romontbos Horizon" and the right-(west-) dipping proto-flint layers above.

defined the genesis of wavy bedded cycles. In a numerical model for wavy bedded Tuffaceous Chalk cycles, sea-level and subsidence rate are presumed to be constant, while the hydrodynamic energy, deposition rates and depth of storm reworking vary periodically. Carbonate cement and/or authigenic silica precipitate at constant rate in a redox zone at constant depth below and parallel to the fair-weather sediment surface and the zone of bioturbative mixing.

In order to simulate the genesis of wavy bedded cycles, the model for planar-parallel bedded sequences (Chapter 5, 6) has to be extended with an extra spatial dimension, allowing the simulation of a series of laterally succeeding sequences. Furthermore, the model must contain rules that define the variation of the wave-length, wave-height and

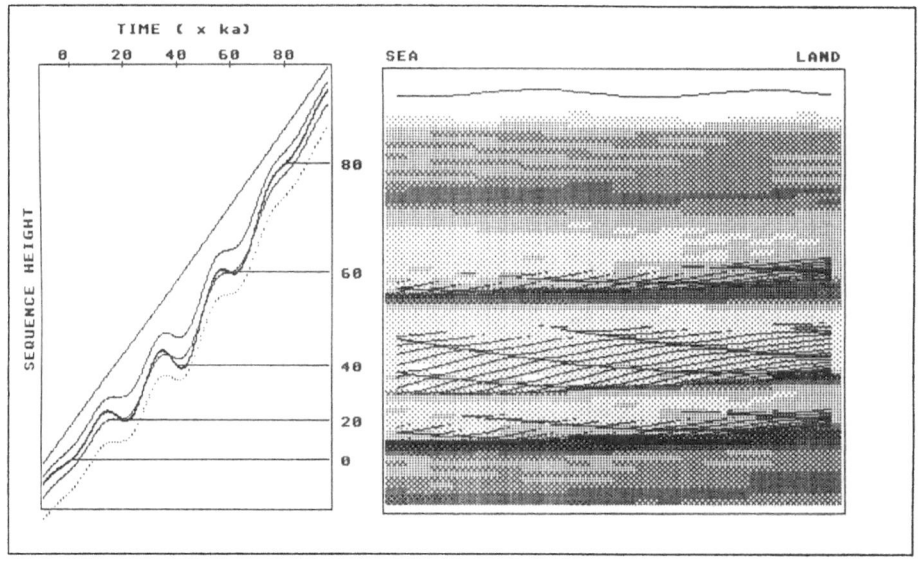

Figure 7.16 - Result of the simulation of the genesis of the wavy bedded middle part of the Gironde sequence (Figs. 7.3, 7.4). Left box: As in Figure 7.15 but for a 100 ka period. Right box: 5 precession cycles of which the middle 3 high energy cycles show storm erosion surfaces, that have been destroyed by bioturbative mixing during the lower energetic periods. The lateral migration rates of the wavy storm erosion surfaces in the sea-ward direction have been constant and low. The wave-height decreased land-ward and was proportional to the periodically varying hydrodynamic energy. Lithification (dark=strong, light=weak) is well developed at the top of the first cycle, formed during the period of lowest deposition rates. The lithified layers have been eroded during the moments of maximum hydrodynamic energy and the depth of erosion was proportional to the maximum storm energy. Note that when maximum storm energy would have been higher, a further increase of the depth of reworking would have been hampered by the previously lithified sediment.

wave-migration rate (celerity) of the wavy fair-weather sediment surface, zones of bioturbation and authigenesis and of the storm erosion surface (see appendix E).

### 7.4.1 Simulation results

The first simulation concerns the wavy bedded cycles with flint layers of the middle part of the Maastricht sequence (Figs. 7.5, 7.6, 7.15). In figure 7.5 one observes at the middle of the sequence at least 4 planar flint layers dipping to the right. These occur in large-scale cross-laminated, fine-grained bioclastic sands, with laminae also dipping to the right. Between the flint layers, barely visible erosion surfaces are present, that also dip towards the right, although somewhat steeper than the flint layers and consequently, the erosion surfaces cut off the flint layers at the right-hand side. Below the right-dipping flint layers, a poorly preserved left-dipping erosion surface and a flint layer below and parallel are present.

The left-dipping erosion surface (Horizon of Romontbos; Felder, 1975a,b) formed during the moment of maximum storm energy of the precession period. The simulation starts at 10 ka before, at the moment of minimum storm energy (Fig. 7.15). Storm reworking depth is then also at a minimum and the deposition rate equals the subsidence rate. The fair-weather sediment surface, the zone of bioturbative mixing, the zone of authigenic silica precipitation and the storm erosion surface are planar-bedded.

Approximately 5 ka later, the deposition rate reaches a minimum, the sediment is reworked by storms to a depth that is still less than the thickness of the zone of bioturbative mixing and the storm erosion surfaces are not yet preserved. The fair-weather sediment surface, the zone of bioturbative mixing, the zone of authigenesis and the storm erosion surface have become wavy and migrate laterally with maximum celerity, while the wave-height of the storm erosion surface is greater than that of the other surfaces.

During the following 5 ka, maximum hydrodynamic energy and depth of storm reworking are reached, while deposition rates increase. During the moment of high hydrodynamic energy, the depth of storm reworking exceeds the thickness of the zone of bioturbation and then the deepest parts of the wavy storm erosion surfaces are locally preserved. While wave-length remains the same, the wave-heights increase and the celerities decrease. During the subsequent decrease of the hydrodynamic energy, wave-heights decrease and celerities increase again.

The simulation results are in accordance with the cycle morphology observed in the field. The simulation has produced left-dipping layers of concentrated authigenic silica, while directly above, a left-dipping erosion surface equivalent to the Romontbos Horizon has formed. The Romontbos Horizon has been formed during the lateral migration to the right of the troughs of the wavy storm erosion surfaces. A succession of right-dipping layers of concentrated authigenic silica is present above this erosion surface and these layers have been cut off at the right-hand side by the troughs of the wavy storm erosion surfaces.

The second simulation concerns the wavy-bedded cycles of the middle part of the Gironde sequence (Figs. 7.3, 7.4, 7.16). The same rules as in the previous simulation have been used. The wave-lengths of the fair-weather sediment surface, zone of bioturbative mixing, zone of carbonate cement precipitation below and of the storm erosion surface remain constant. The wave-heights are proportional to the periodically varying hydrodynamic energy and the depth of storm reworking. Additionally, wave-heights increase towards the left, in accordance with a presumed increase of wave-height with increase of depth in the seaward direction. The celerities of the waves are again presumed to have been inversely proportional to the hydrodynamic energy and the wave-heights.

Also this second simulation of the genesis of a 100 ka eccentricity cycle with 5 precession cycles, resembles the field sequence. At the basis of the field and simulated sequences one observes relatively thin, homogeneously bioturbated and well lithified tops of cycles, ending with a planar erosion surface and covered by relatively steep, laterally succeeding wavy storm erosion surfaces. At the top of the field and simulated sequences, cycles are thicker, storm erosion surfaces at the basis of the cycles are poorly preserved, dip more gently and succeed each other more closely. The major part of the upper cycles has been homogeneously bioturbated and lithification is less well developed than in the lower cycles.

## Discussion

The rhythmic variation of bed thickness, grainsize, structures and authigenic mineral concentrations in the planar-parallel bedded Chalk shows a regularity that has been attributed to the influence of periodic variations of the Earth's orbital parameters, insolation, climate and oceanography (Hart, 1987; Cottle, 1989; Herrington et al., 1991; Chap. 5).

Elsewhere, the existence of Milankovitch cycles has been proved to a certain extent, by frequency analysis of long sequences deposited at constant rates and with a variation of lithology that reflects a more or less direct response to the periodic variations of the Earth's orbital parameters (Imbrie & Imbrie, 1979; De Boer, 1983, 1991a,b; Fisher, 1991; and many others).

It has been argued (Chapter 5) that Chalk sequences with flint nodule layers are reflecting the variation of the precession index and that by using a proper model of the depositional-early diagenetic environment in reverse, the field sequence can in principle be used to reconstruct the precession index.

Because the parallel-bedded Chalk sequences of Maastricht and the Gironde Estuary gradually change upwards into wavy-bedded Tuffaceous Chalk sequences, it is assumed that the cycles of the much more irregularly bedded Tuffaceous Chalk sequences have also been formed during precession periods. It has been shown that when the depositional-early diagenetic model for planar-parallel bedded Chalk (Chapter 5, 6) is extended with rules that define the relation between the periodic variation of the hydrodynamic energy and the wave-lengths, wave-heights and celerities of the fair-weather sediment surface, the zone of bioturbative mixing, the redox zone of mineral authigenesis and the storm erosion surface, also the genesis of wavy bedded precession cycles can be simulated and clarified.

In (Tuffaceous) Chalk with a low detrital skeletal opal concentration, lithification by carbonate cement in the anoxic redox zones is particularly strong during periods of increasing average hydrodynamic energy and low deposition rates. When during the subsequent period of high hydrodynamic energy and deep storm reworking the lithified layer is eroded, then wavy bedding is generated. The further lithification of the sediment in a redox zone below and parallel to the wavy fair-weather sediment surface decreases the erodability and preserves the wave morphology, that changes only gradually during subsequent storm reworking. The complex feed-back relation between storm reworking and lithification has to be further investigated (Chapter 8, 9).

# Conclusion

The Maastrichtian (Tuffaceous) Chalk sequences of Maastricht and the Gironde Estuary have been deposited in a subtropical shallow marine epi-continental sea. The upward coarsening of the grainsize and the increase of the average bed thickness indicate a gradual increase of the average hydrodynamic energy and deposition rates. The rhythmic vertical variation of the bed thickness, grain-size, structures and authigenic mineral concentrations reflect a superimposed periodic variation of hydrodynamics and deposition rates. The periodic variation of depositional and early diagenetic conditions is presumed to reflect the variation of the precession index.

Silica and/or carbonate cement precipitated in (Tuffaceous) Chalk at some depth below the sediment surface, in the anoxic zones of sulphate and/or carbondioxide reduction. Most strongly silicified/lithified layers formed when the deposition rate was zero, thus when hydrodynamic energy increased and the consequent increase of erosion was equal to the relative sea level rise. During a further increase of the hydrodynamic energy, previously lithified sediment was eroded during storms and wavy beds formed. A hardground, a bored, encrusted and mineralised rocky sea bottom, formed at relatively high average hydrodynamic energy when the sediment that was eroded during a storm, was no longer redeposited after the storm, so that the previously lithified layer was continuously exposed.

After extending the above developed numerical models for the simulation of the genesis of rhythmic bedding in parallel-bedded Chalk (Chapter 5, 6) with rules that define the relation between wave-heights, wave-lengths and celerities of the wavy fair-weather sediment surface, wavy zone of bioturbative mixing, wavy redox zone of silica and/or carbonate cement authigenesis and the wavy storm erosion surface, the genesis of wavy bedded Tuffaceous Chalk cycles could also be simulated.

It appears that realistic results can be obtained if the wave-heights of the fair-weather sediment surface, the zone of bioturbation and the zone of authigenesis are the same and smaller than the wave-height of the storm erosion surface. The wave-lengths are constant, while the wave-heights are proportional and the celerities are inversely proportional to the periodically varying hydrodynamic energy.

During the migration of the sediment waves, oblique, lithified layers may form at the wave flanks that dip oppositely to the direction of migration and that are characterised by lowest deposition rates. The lithified layers hamper further erosion and eventually lithification influences the wave morphology and the hydrodynamics above the wavy sea bottom. There exists a feed-back between lithification and depth of storm erosion.

## Appendix E - Numerical model

The numerical model for the genesis of parallel bedding is extended with rules that define the wavelength, wave height, and wave migration rate of the fair-weather sediment surface (FS), zones of bioturbation (BD) and authigenesis (RD) and storm erosion surface (SS). All wave parameters are the same, except the wave height of the storm erosion surface that is a constant factor larger than those of the other surfaces and zones. The position of the surfaces in time (Fs(t), Bd(t), Rdmin(t), Rdmax(t) and Ss(t)) as a function of the subsidence rate and the periodically varying hydrodynamic energy (E(t)), and the processes of bioturbation and authigenesis below the wavy fair-weather surface (Fs(i,t)) are defined according to the procedures described in appendix C (Chapter 6).

```
+ = addition
- = substraction
* = multiplication
/ = division
pi=3.14..
sin = goniometric function
A(m,n) = array with m x n sites

FOR t=0 TO t=T
    FOR i=0 TO i=I
            Fs(i,t)=Fs(t) + Amax*(DFS/DSE)*E(t)*sin(((2*pi)/L)*i + (f1-E(t))*f2*(t/T)*(2*pi))
            Bd(i,t)=Fs(i,t) - Bd
            Rdmin(i,t)=Fs(i,t) - Rdmin
            Rdmax(i,t)=Fs(i,t) - Rdmax
            Ss(i,t)=Ss(t) + Amax*E(t)*sin(((2*pi)/L)*i + (f1-E(t))*f2*(t/T)*(2*pi))
    NEXT i
NEXT t
```

T = simulation time
I = width of the simulation space
Fs(i,t) = vertical position of the fair-weather sediment surface at i during t
Fs(t) = average vertical position of the fair-weather sediment surface during t
Bd(i,t); Bd(t); Bd = bioturbation depth below the fair-weather sediment surface
Rdmin(i,t); Rdmin; Rdmax(i,t); Rdmax = minimum and maximum depth of the zone of authigenesis below the fair-weather sediment surface
Ss(i,t); Ss(t) = storm erosion surface
Amax = maximum amplitude
DFS = depth of fair-weather sediment surface below sea level
DSE = depth of storm erosion surface below sea level   DSE>DFS
E(t) = periodically varying hydrodynamic energy 0<E(t)<1
L = wave length
f1; f2 = proportionality factors f1>1 f2>0

# 8  Genesis of the K/T Section of Stevns Klint

### Abstract

**A numerical model that was developed for the simulation of the genesis of Maastrichtian wavy-bedded (Tuffaceous) Chalk precession cycles of Maastricht (SE Netherlands) and the Gironde Estuary (SW France), is used to explain the sedimentology of the wavy-bedded Maastrichtian-Danian (Tuffaceous) Chalk sequence of Stevns Klint (Denmark). It appears that the Stevns Klint K/T boundary Fish Clay with an abnormally high iridium concentration reflects the dissolution of carbonate and the concentration of insoluble authigenic minerals (condensation) during a period of exceptionally low deposition rates, that presumably coincided with a rather low-energetic precession period, during a sudden change of tectonic regime.**

## Introduction

The Stevns Klint cliff section is exposed along the east coast of Sjaelland (Denmark), 50 km south of Copenhagen. It is part of the Norwegian-Danian basin, situated between the Ring-Købing-Fyn High and the Fenno-Scandinavian Shield (Rasmussen, 1978; Surlyk, 1979).

The sequence is a classical, 40 m high exposure (Abildgaard, 1759). Its lower half consists of Maastrichtian (Dumont, 1849) white coccolithic mudstones with (sub)horizontal, planar and undulating flint nodule layers (White Chalk), and its upper half of Danian (Desor, 1846) coccolithic-bryozoan wacke-/packstones with undulating flint nodule layers (Bryozoan Limestone).

The boundary between the Late Cretaceous Chalk and the Early Tertiary Limestone (Tuffaceous Chalk) is complex and its lithostratigraphy was first properly interpreted by Ravn (1925) and was later studied in detail by Christensen et al. (1973). The surface of the Maastrichtian Chalk is wavy and in the troughs covered by a cm thin pyrite layer followed by smectitic clay with fish remains and abnormally high iridium concentrations

144

Figure 8.1 - Cretaceous/Tertiary boundary sequence (30 m) exposed at Højerup (Stevns Klint, Denmark) (for scale and description see Fig. 8.2).

(Fish Clay).

Recently the Fish Clay has received considerable attention because of a hypothesis that explains a presumed (Voigt, 1979b) mass extinction at the end of the Cretaceous as the result of a catastrophic impact of an extraterrestrial iridium-rich body (Alvarez et al, 1980; Smit & Hertogen, 1980). The highest measured iridium concentration (185 ppb) is several orders of magnitude higher than the values normally measured in Chalk and Limestone (Hansen et al, 1986). The absence of calcareous plankton and anomalously high concentrations of Co, Ni, Cr and Ir in samples of the Fish Clay (Smit & ten Kate, 1982; Hansen et al., 1986) have been considered to support the impact hypothesis.

Because not only the Ir concentration is high but also the concentrations of clay minerals, pyrite and organic matter while at the same time the carbonate concentration is low, dissolution of carbonate and condensation of insoluble residue might as well have caused the concentration of Ir and the absence of calcareous plankton (Rampino, 1982; McLean, 1985).

It appears necessary to understand the genesis of the Stevns Klint section before meaning can be assigned to the measured high Ir concentration. Several sedimentological aspects of the Stevns Klint sequence will be discussed. A numerical model for the deposition

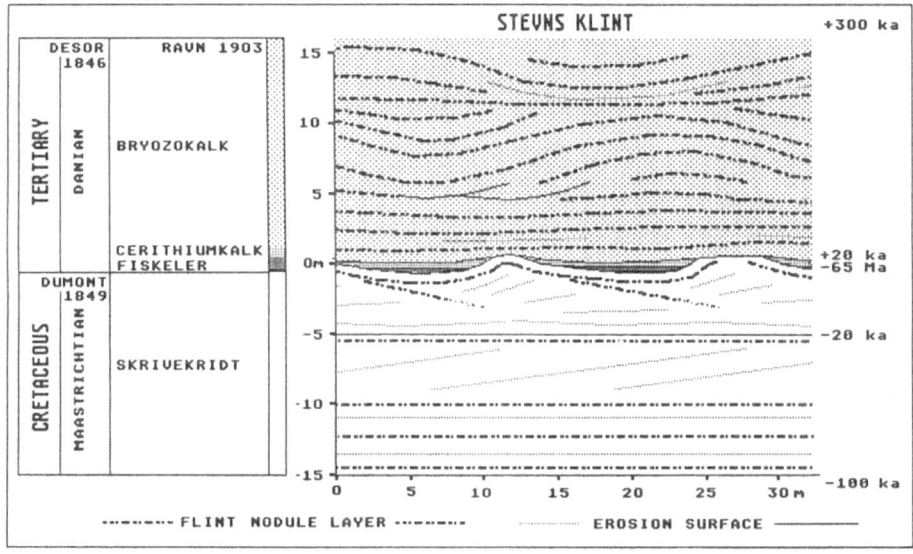

Figure 8.2 - Schematic representation of the lithostratigraphy of the K/T boundary sequence at Højerup (Stevns Klint, Denmark). Lower 15 m consists of Maastrichtian coccolithic mudstones (White Chalk) with planar-parallel, oblique and undulating flint nodule layers and erosion surfaces. The undulating K/T boundary surface is covered in the troughs by the iridium-rich smectite of the Danian Fish Clay, that gradually changes upwards into the Cerithium Limestone. The upper 15 m consist of Danian coccolithic bryozoan wacke-/packstones (Bryozoan Limestone) with undulating flint nodule layers and erosion surfaces.

and the early diagenesis of Chalk, developed for the simulation of the genesis of bedding, as it was observed in the Maastrichtian Chalk of Maastricht (The Netherlands) and the Gironde Estuary (France), is used to simulate the genesis of the K/T boundary sequence at Stevns Klint.

# 8.1 Litho-/biofacies

### 8.1.1 The Maastrichtian White Chalk

The Maastrichtian White Chalk exposed at Højerup can be divided into 3 intervals (Figs. 8.1, 8.2). In the lower third of the sequence of Maastrichtian Chalk 3 slightly cemented beds with rather poorly developed planar-parallel flint nodule layers are exposed. The

146

Figure 8.3 - K/T boundary at Højerup (Stevns Klint, Denmark). The trough in the Maastrichtian White Chalk has been filled with Danian Fish Clay, that gradually changes upwards into the Cerithium Limestone. Eroded Cerithium Limestone and Maastrichtian Chalk are covered by Danian Bryozoan Limestone (for scale and description see Fig. 8.4).

middle third of the White Chalk is formed by a thick bed with a nodularly cemented and eroded top, with directly below, a well developed planar flint nodule layer. The upper third of the sequence is formed by a thick Chalk bed that is characterised by oblique flint nodule layers, laterally succeeding each other at regular distance and upwards amalgamating with an undulating flint nodule layer, 0.5 m below and parallel to the undulating K/T boundary.

The Maastrichtian Chalk is, besides flint nodule layers, also characterised by (sub) horizontal erosion surfaces, that are planar in the lower part and oblique in the middle and upper part. In the upper part erosion surfaces dip in a direction opposite to the direction of the dip of the oblique flint nodule layers (Figs. 8.1, 8.2).

## 8.1.2 The Danian Fish Clay and Cerithium Limestone

The troughs upon the undulating top of the Maastrichtian Chalk (Figs 8.1-8.4) are covered

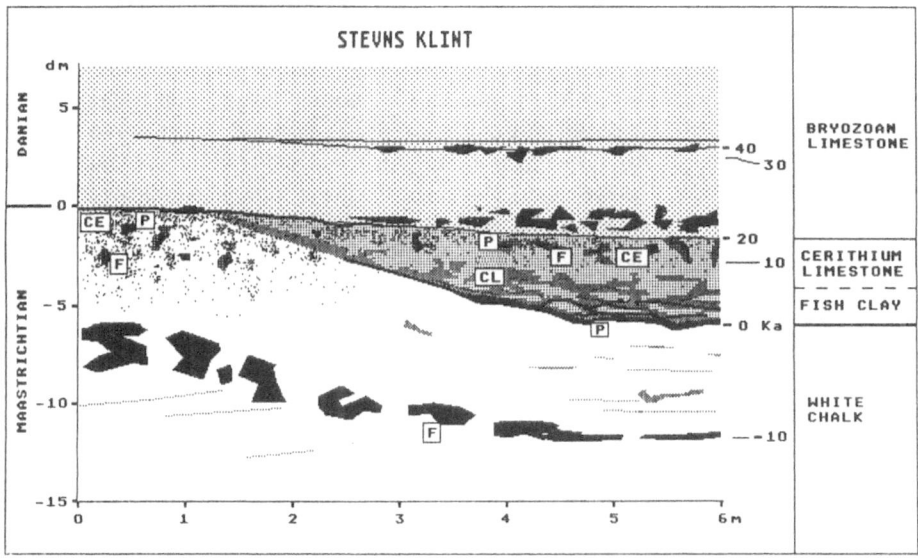

Figure 8.4 - K/T boundary at Højerup (Stevns Klint, Denmark). CL (smectitic clay), P (oxidized pyrite), CE (carbonate cement), F (Flint). Times of highest authigenic mineral concentration in layers during minimum deposition rates, and of storm erosion surfaces (fat lines) at maximum storm intensity are indicated (text).

by a cm-thick organic-rich layer with pyrite nodules (Christensen et al., 1973). This layer wedges out towards the eroded crests of the Maastrichtian Chalk. Locally a few cm of laminated smectitic Chalk occur below the pyritic layer. Above the pyritic layer, dark-coloured, laminated, organic-rich smectite (Fish Clay) changes gradually upwards into a grey calcareous smectite with compacted lumps of more or less smectitic Chalk. The clay mineral, organic matter and iridium concentrations are highest just above the pyrite layer that covers the Maastrichtian Chalk in the troughs. The concentrations decrease upwards, while the carbonate content increases simultaneously (Hansen et al., 1986).

The nodular smectitic Chalk above the pure smectite of the Fish Clay changes gradually upwards into a pure carbonate wackestone with molds of the gastropod *Cerithium* (Cerithium Limestone). The Cerithium Limestone has been cemented and brecciated. Silica concretions are poorly developed and have formed around *Thalassinoides* burrows that have been enlarged due to dissolution and/or erosion (Bromley, 1968). *Thalassinoides* burrows may have iron-oxide-coated walls (late-diagenetically oxidized pyrite). Pyritization, lithification and silicification are not restricted to the Cerithium Limestone in the troughs, but have also affected the crests of the Maastrichtian Chalk in between. The Cerithium Limestone and the Maastrichtian Chalk have been partly eroded and the erosion surface is smooth and slightly undulating.

### 8.1.3 The Danian Bryozoan Limestone

The Danian Bryozoan Limestone (Figs 8.1-8.4) is a pack-wackestone of delicate bryozoan debris with a fine-grained matrix of coccoliths and planktonic foraminifera (Thomsen, 1976). The bedding is defined by undulating flint nodule layers. The wave-length of the flint layers is rather constant in the vertical succession, while the wave-height is variable (Thomsen, 1976). One may distinguish 3 lithostratigraphic intervals of approximately 5 layers. The flint nodule layers of the lower interval have low amplitudes and are followed by a middle interval with high amplitudes. The youngest interval has nodule layers with an intermediate amplitude. Between and parallel to the flint nodule layers erosion surfaces occur that form, in particular in the troughs, top-surfaces of underlying eroded cycles with a more or less well developed hardground.

### 8.1.4 Lateral changes in the Stevns Klint section

Several kilometres towards the southwest (Rødvig, Stevns Klint), the wavy character of the succession disappears. The upper 5 flint nodule layers of the Maastrichtian are planar, they are of similar thickness and the whole interval is thinner than in the northern exposures. Older Maastrichtian with undulating flint nodule layers crops out below.

At Rødvig, the K/T boundary is defined by a gently undulating, diffuse boundary of somewhat more smectitic Chalk, without a pyrite layer, and with only a poorly developed cementation of the Cerithium Limestone above. The Danian Bryozoan Limestone is also characterized by only gently undulating flint nodule layers.

## 8.2 Summary of the Sedimentology of Chalk

The Stevns Klint carbonates are part of the Maastrichtian-Danian (Tuffaceous) Chalk of NW Europe that has been deposited in an extensive shallow marine epi-continental sea (Håkansson et al., 1974). Several tens of metres thick sequences, exposed near Maastricht (The Netherlands) and along the Gironde Estuary (France), have been studied and some aspects of their genesis are summarized below.

The sequences are characterised by a gradual coarsening-upward of grain size and thickening-upward of dm- to m-thick beds. Beds are laterally continuous over distances of at least several km (Felder, 1975a,b) and they are defined by a rhythmical vertical variation of grainsize (Felder, 1986), depositional-bioturbational structures (Ekdale & Bromley, 1991; Savrda et al., 1991) and authigenic mineral concentrations (Clayton, 1986; Chapter 6).

The deposition rates of (Tuffaceous) Chalk have been in the order of several cm-dm/ka (Håkansson et al, 1974). A regular variation of bed thickness and lithology has been interpreted to reflect the influence of the precession index on climate, oceanography and depositional-early diagenetic processes (Hart, 1987; Cottle, 1989; Gale, 1989; Leary et al., 1989; Herrington et al., 1991; Chapter 5).

The (Tuffaceous) Chalk has a very low content of terrigenous siliciclastics and consists almost entirely of calcareous bioclasts with up to 20% of silica, concentrated in pure cryptocrystalline quartz concretions (flint, Buurman & van der Plas, 1971), derived from late-diagenetically dissolved opaline bioclasts (i.e., diatoms, radiolaria and sponge spicules, Soudry et al., 1981). Several percent of the rock volume is formed by early-diagenetic authigenic minerals, that precipitated as a result of bacterial decomposition of organic matter (Chapter 3). Authigenic (potassium) iron-phyllosilicates (glauconitic smectite) (Burst, 1958; Harder, 1980; Odin & Matter, 1981), iron-sulphides (Roy & Trudinger, 1971; Pyzic & Sommer, 1981; Drobner et al., 1990), carbonate cement (Hudson, 1977; Gautier & Claypool, 1984; Raiswell, 1987) and silica (Siever, 1962; Zijlstra, 1987) formed in the above order in aerobic to anoxic redox zones around deep burrows (Bromley et al. 1975; Clayton, 1986; van der Weijden et al., 1989; Zijlstra, 1989) and below and parallel to the seabottom (Redfield et al., 1963; Froelich et al., 1979, Berner, 1980; Chapter 6).

The more than a metre thick cycles in the bioclastic silt-/sandstones of the Tuffaceous Chalk are asymmetric truncated (Einsele et al., 1991) fining-upwards sequences. The basis of the cycles consists of an undulating erosion surface, covered by phosphatic-glauconitic-pyritic, coarse-grained and wavy-laminated, bioclastic sandstone, changing upwards into a homogeneously bioturbated, bioclastic siltstone, that is lithified and moreover eroded, bored and encrusted (hardground; Voigt, 1929; 1968, 1974). The fining-upwards cycles are considered to be tempestite cycles, deposited during approximately 20 ka precession periods (Chapter 5, 6, 7). They reflect the increase-decrease of average storm frequency and intensity, presumably the result of a latitudinal shift of the caloric equator and associated climate zones. During the increase of the storm intensity sediment was eroded and during the decrease of the storm intensity a multi-event fining-upwards sequence of amalgamated tempestites was deposited.

The dm-m thick cycles in the fine-grained Chalk are homogeneously bioturbated symmetric cycles with a much less pronounced grainsize variation. These are also considered tempestite precession cycles. However, the hydrodynamic energy of the depositional environment was rather low and consequently storm reworking was shallower than the zone of bioturbative mixing and not only the upper part, but the entire redeposited fining-upwards storm layers were destroyed during reestablished post-storm bioturbation.

The dynamics of the depositional-early diagenetic processes that led to the genesis of a rhythmically bedded Chalk sequence at Stevns Klint can be simulated with the help of numerical models that have been developed for the simulation of the genesis of wavy bedded precession cycles in the Maastrichtian of Maastricht (SE Netherlands) and the Gironde Estuary (SE France) (Chapters 5, 6 and 7). These models are based on the premises that the deposition rates $(dS/dt)$ varied periodically around a long term average $(Sub(t))$ as a function of periodically varying storm intensity $(E(t))$. The variation of the authigenic mineral concentration $(dC/dt)$ is considered to be proportional to the reaction rate $(R)$ and inversely proportional to the deposition rate:

$$dS/dt = Sub(t) - F*dE/dt$$
$$dC/dt = R/(1 + ABS(dS/dt))$$
$$F = proportionality\ factor$$

In those cases where (Tuffaceous) Chalk had a low skeletal opal concentration, carbonate cement, instead of silica, precipitated in the deepest anoxic zones of sulphate and carbondioxide reduction (Chapters 4, 5). In particular when the storm intensity increased and F*dE/dt became equal to Sub(t) so that dS/dt was zero and dC/dt was at a maximum, a well lithified layer formed. During a subsequent further increase of the storm intensity, the top of the lithified layer was repeatedly eroded and covered by a redeposited fining-upward storm layer or even became continuously exposed and developed towards a bored and encrusted hardground.

Wavy storm erosion surfaces, like they were observed in the Tuffaceous Chalk of Maastricht and the Gironde Estuary, formed during relatively deep storm reworking, when the sediment that lithified in the deeper anoxic redox zones was uncovered and eroded. The wavy storm erosion surface was covered again during waning of the storm, resulting in a post-storm fair-weather sediment surface, which wave-height was less than that of the storm erosion surface below. As a result of the wave form of the fair-weather sediment surface, sediment lithified in a wavy anoxic redox zone below and parallel to the fair-weather sediment surface.

During the next storm, the new wavy storm erosion surface followed the top of the wavy lithified layer and was more or less similar to the former storm erosion surface. The feed-back between storm erosion and lithification caused a gradual migration and change of morphology of the wavy storm erosion surface, the fair-weather sediment surface, the zone of bioturbative mixing and the redox zones of mineral authigenesis, as a function of the change of storm intensity and deposition rates during a precession cycle (Chapter 7).

## 8.3  Sedimentary conditions during the K/T transition

The Stevns Klint section is interpreted according to the depositional and early diagenetic model for the genesis of the wavy bedded (Tuffaceous) Chalk sequences of Maastricht and the Gironde Estuary (Chapter 7). The (sub)horizontal erosion surfaces in the Stevns Klint sequence are thus considered storm erosion surfaces that were preserved during times and at sites of relatively high hydrodynamic energy, when the depth of reworking below the fair-weather sediment surface exceeded the depth of bioturbative mixing after redeposition of the fining-upwards storm layer. The flint nodule layers of the Stevns Klint sequence occur in sediment that resided relatively long in the deepest anoxic redox zones of silicification, during times and at sites of relatively low deposition rates.

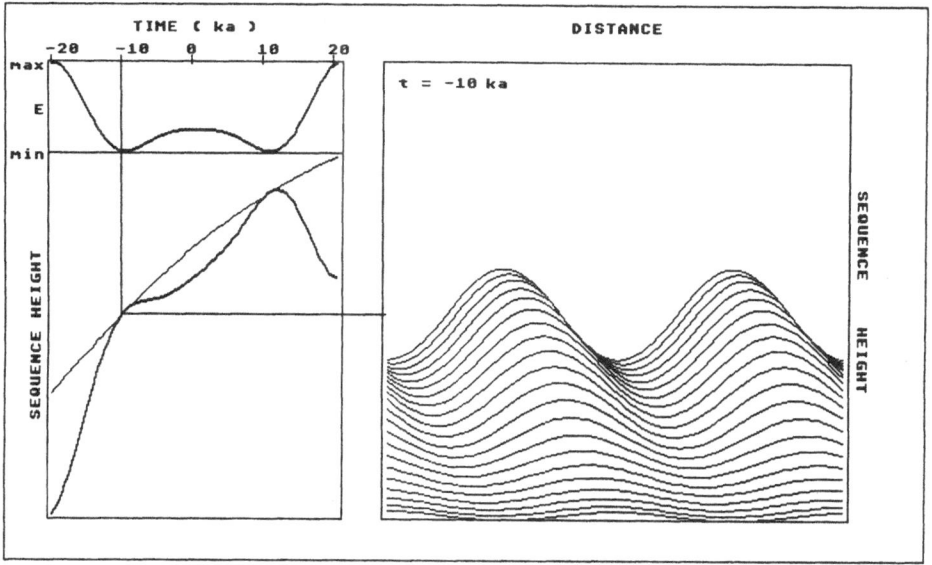

Figure 8.5 - Depositional conditions during 40 ka around the K/T boundary (0 ka). Upper left box: variation of storm intensity. Lower left box: decrease of the average subsidence/deposition rate with superimposed a periodic variation of hydrodynamic energy (depth of storm reworking) and deposition rates. Right box: after sediment had been eroded above the uppermost planar Maastrichtian flint nodule layer that formed during -30 to -20 ka, the depth of storm reworking decreased and wavy storm erosion surfaces (500 year intervals) were formed during -20 to -10 ka. Oblique flint nodule layers formed below the flanks of the wavy erosion surface that dipped opposite to the direction of wave migration, and in sediment that was characterised by a close vertical succession of storm erosion surfaces and low deposition rates.

The 15 m of Maastrichtian White Chalk exposed at Højerup contains 5 laterally continuous flint nodule layers. The 5 flint layers thus indicate times of slow deposition as hydrodynamic energy increased during each of the 5 subsequent precession periods. The erosion surfaces in between have formed during moments of maximum storm intensity and they are used to define the cycle- and precession period-boundaries. The sequence as a whole represents an approximately 100 ka asymmetric eccentricity cycle and average deposition rates have been 15 cm/ka (see also figures 6.10a,b and 6.11a,b).

The variation of the distance between the flint nodule layers and storm erosion surfaces reflects the variation of the maximum storm intensity, hydrodynamic energy, deposition rates and depth of storm reworking during the 5 subsequent precession periods. For instance, the relatively thin cycles in the lower third of the succession indicate deposition during relatively low energetic precession periods. Storm reworking was shallow during times of maximum, but low storm intensity and consequently the deeper anoxic sediment

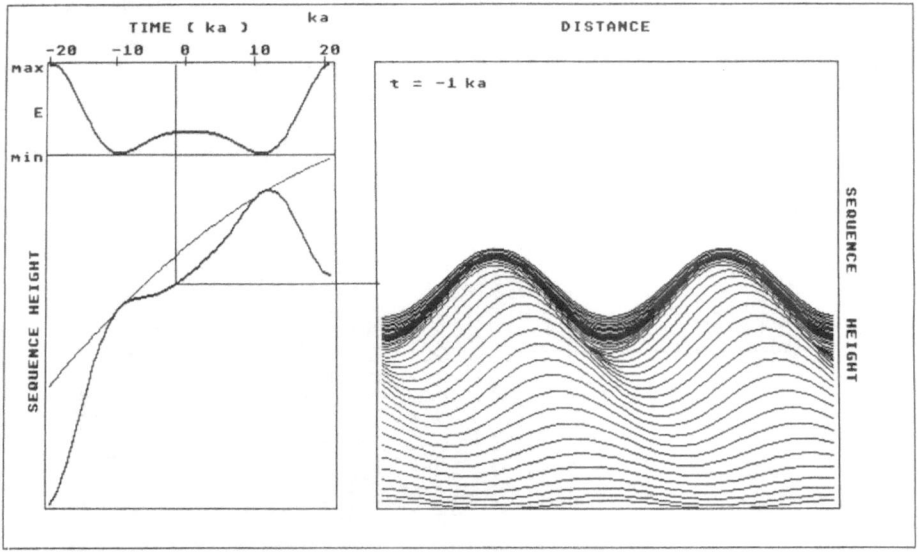

Figure 8.6 - For explanation see Fig. 8.5. The uppermost Maastrichtian undulating flint layer (Figs. 8.2,
8.4) has been formed below the undulating K/T boundary during -10 to 0 ka. The condensed clay deposit
and pyrite layer formed during the same period of increasing hydrodynamic energy and lowest deposition
rates, followed by the deposition of Cerithium Limestone at increasing deposition rates and again decreasing
depth of storm reworking during 0 to 10 ka (Fig. 8.7).

in which the (proto) flint nodule layers had formed during the earlier increase of storm
intensity and during minimum deposition rates, was not reworked. The storm erosion
surfaces typically occur at some distance above the flint nodule layers.

To the contrary, the thick uppermost Maastrichtian precession cycle reflects a high
storm intensity maximum. Deep reworking is indicated by the fact that the nodularly
lithified Chalk and planar flint nodule layer, 5 m below the undulating top of the cycle
(K/T boundary) (Fig. 8.2), is directly followed by the storm erosion surface that formed
at the moment of maximum storm energy, approximately 20 ka before the genesis of
the undulating K/T boundary erosion surface. The dynamics of the genesis of the Stevns
Klint K/T boundary sequence is discussed and simulated (figs 8.5, 8.6., 8.7) for a 40
ka time-span, containing the last Maastrichtian and first Danian precession period.

**-20 to -10 ka** - After the erosion surface formed during maximum storm intensity,
energy decreased again and deposition rates reached their maximum at -15 ka, while
5 m of Chalk was deposited during -20 to -10 ka at an average rate of 50 cm/ka (Fig.
8.5). The oblique erosion surfaces and oppositely dipping oblique flint nodule layers
in the uppermost Maastrichtian precession cycle reflect the genesis and lateral migration
of wavy sediment surfaces, storm erosion surfaces and zones of bioturbative mixing
and authigenic silica precipitation, during the decrease of the storm intensity and the

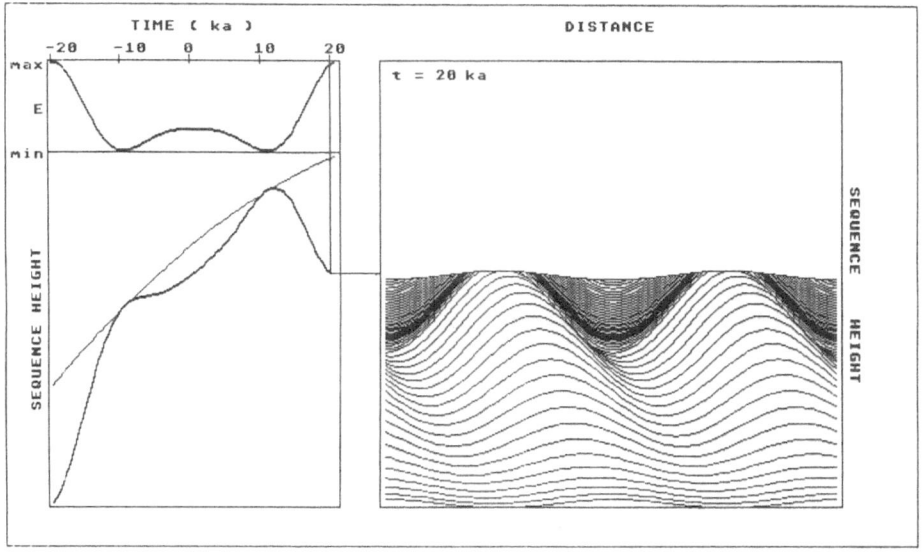

Figure 8.7 - For explanation see Fig. 8.5. Lithification and silicification of the Cerithium Limestone and of the crests of Maastrichtian Chalk in between occurred between 10 and 20 ka when hydrodynamic energy increased again and deposition rates were at a minimum. Both deposits were partially eroded as hydrodynamic energy reached its maximum at 20 ka.

increase of deposition rates (-20 to -10 ka). The climbing ripple-like bedding was characterised by a spatial variation of deposition rates, that was superimposed on the temporal variation. Lowest deposition rates and highest silica concentrations (oblique flint layers) occurred at the wave-flanks that dipped opposite to the direction of wave migration (Ch. 7).

**-10 to 0 ka** - Below the undulating top (K/T boundary) of the thick uppermost Maastrichtian precession cycle (Fig. 8.2), an undulating (proto) flint nodule layer formed during the last 10 ka of the Maastrichtian, when hydrodynamic energy increased again and deposition rates reached a minimum. Because the uppermost Maastrichtian precession cycle is extraordinarily thick and hardly eroded, and because the storm erosion surface of the K/T boundary (0 ka) occurs at some distance above the undulating flint nodule layer, much like the lowermost low-energetic Maastrichtian precession cycles, it is suggested that the first Danian precession period (0 to 20 ka) was rather low energetic (Fig. 8.6).

During the last 10 ka of the Maastrichtian, the storm intensity increased and deposition rate decreased. Sediment was repeatedly reworked during storms and redeposited on the undulating top of the Maastrichtian Chalk. The continuous reworking and oxygenation of the sediment caused the oxidation of organic matter, the dissolution of carbonate and the increase of the concentration of insoluble residue (Chapter 6). After reworking and oxygenation, the dissolved oxygen concentration in the pore fluid of the redeposited sediment decreased again and authigenic smectite formed under slightly reducing conditions.

At the undulating storm erosion surface, above the anoxic Maastrichtian Chalk characterised by sulphate reduction, and below the redeposited smectitic Chalk characterised by iron-oxide reduction, reduced iron reacted with reduced sulphur in order to precipitate as iron-sulphides (pyrite). Because the silicified/lithified undulating top of the Maastrichtian Chalk resisted erosion, depth of reworking hardly increased despite the increase of hydrodynamic energy during the last 10 ka of the Maastrichtian. Consequently, a relatively thin layer of sediment was repeatedly reworked and the alternating oxygenation and mild reduction of the same sediment in the sub-oxic zone above the stationary storm erosion surface caused the exceptional dissolution of carbonate, concentration of iridium and genesis of the smectite in the condensed Fish Clay.

**0 to 10 ka** - When the storm intensity and hydrodynamic energy decreased again during the first 10 ka of the Danian, the depth of storm reworking decreased and the deposition rates increased. The increase of deposition rates resulted in the decrease of the condensation of insoluble residue and the net increase of detrital carbonate deposition. As a result of the decrease of storm reworking depth, depositional lamination was destroyed during bioturbative mixing in the Cerithium Limestone.

**10 to 20 ka** - During the second half of the Danian precession period (10 to 20 ka), when the storm intensity increased again and the deposition rates decreased, a flint layer formed and carbonate cement precipitated in the anoxic upper part of the Cerithium Limestone and in the crests of the Maastrichtian Chalk in between. During maximum hydrodynamic energy and depth of reworking (20 ka) part of the previously lithified White Chalk and Cerithium limestone were eroded (Fig. 8.7).

The lower 5 wavy flint nodule layers of the Bryozoan Limestone were formed during the first Danian eccentricity period. They are also inferred to reflect the periodic variation of the precession index and storm intensity, although the asymmetry of the eccentricity cycle is obscured by the wavyness of the flint nodule layers and storm erosion surfaces. The wavyness is preserved and inherited during the deposition of the subsequent precession cycles. Furthermore, the larger wave-length and in particular the larger wave-height have resulted in lower lateral migration rates of the wavy fair-weather sediment surfaces, storm erosion surfaces and zones of bioturbation and authigenesis during the Danian, as compared to the last wavy-bedded Maastrichtian precession cycle.

## Discussion

The reconstruction of the variation of the depositional and early diagenetic conditions during deposition of the Maastrichtian-Danian sequence at Stevns Klint suggests that the high iridium concentration in the Fish Clay at the K/T boundary is the result of a decrease of the net deposition rate and of the concentration of Ir in a favourable redox zone.

The Maastrichtian Chalk was deposited at an average rate of 15 cm/ka. The truncated and eroded precession-induced sedimentary cycle that is represented by the Fish Clay and the Cerithium Limestone is up to 60 cm thick at the centre of the troughs in the Maastrichtian Chalk, and has been deposited during approximately 10 ka. The average deposition rate was thus of the order of only 3 cm/ka. Because the deposition rates decreased and subsequently increased during the precession period, it is plausible that during the increase of hydrodynamic energy, deposition rates were locally zero during a period of several ka. The low deposition rates caused an increased residence of sediment in the various redox zones of bacterial metabolism. It was shown (Colodner et al., 1992) that Ir is soluble in a reducing environment and that it co-precipitates with Mn and/or Fe oxyhydroxides in a sub-oxic environment. The highest Ir concentrations (185 ppb) have indeed been measured in the smectite at the basis of the Fish Clay (Hansen, 1986), which presumably formed in the sub-oxic zone of manganese oxide reduction and clay mineral genesis, situated, during several ka, above the undulating erosion surface and anoxic zone of sulphate reduction and pyrite genesis, at the basis of the zone of continuous storm reworking.

The wavy bedding caused periodic spatial variations of deposition rates, superimposed on the temporal variation as a function of the precession index. High Ir concentrations in the Højerup sequence are thus partly related to the wavy bedding, the more because the smectite and Ir concentrations are much lower in the poorly developed Fish Clay and less wavy bedded sequences (Hansen, 1986), like the Rødvig sequence situated a few km to the South.

The genesis of the wavy bedding in the Maastrichtian Chalk and the Danian Bryozoan Limestone is not well understood. Although bryozoa are common, it has been doubted (Rasmussen, 1971) whether the growth of mound structures during depositional phases was a result of baffling by bryozoans only (Thomsen, 1976). The prominent lamination, sediment sorting, erosion surfaces and lithification in the bryozoan sands of Stevns Klint, the absence of the remains of baffling organisms in similar mound structures in Turonian-Santonian Chalk of Haute Normandie (France; Kennedy & Juignet, 1974) and preliminary results of computer simulations (Chapter 7, 9) suggest that wavy bedding could have been caused by the feedback between storm erosion and lithification as well.

The conditions that led to the wavy bedding and the concentration of Ir at the K/T boundary have been exceptional and local. This is also indicated by the extreme thickness of the uppermost beds of the Maastrichtian at Højerup in comparison with the bedding at Rødvig. A similar local thickening is also characteristic of the precession-induced cycles of the Zumaya sequence (Spain; ten Kate & Sprenger, 1993) and of the uppermost precession-induced cycles of the Maastrichtian (Meerssen Member, Felder, 1975a,b) in the type locality (Maastricht, The Netherlands). Recently, also in this area a Maastrichtian/Danian precession cycle with well preserved smectitic clay layers was found on top of a thick Maastrichtian precession cycle, exposed in a subterranean quarry near quarry Curfs (Geulhem), while elsewhere uppermost Maastrichtian cycles are relatively thin and a smectitic Danian precession cycle is absent. In the K/T boundary clay layers of the Maastricht area, exceptionally high iridium concentrations have not been detected (pers. comm. Smit). In the coarse-grained Tuffaceous Chalk of Maastricht, clay layers occur typically at the top of the storm fining-upwards sequences, well above the storm erosion surface and the eroded Maastrichtian Tuffaceous Chalk below. Presumably, the clay layers remained most of the time above the sub-oxic redox zone where concentration of iridium could occur.

It is not yet clear what has caused the local thickening of the cycles at the end of the Maastrichtian. It could have been caused by a sudden local increase of the subsidence rate related to increased tectonic activity in relation to the massive Indian Deccan Traps volcanism (McLean, 1985).

Although the high Ir concentration in the K/T boundary clay can be explained with the above model to be the result of condensation and concentration in a favourable redox zone during a period of low deposition rates, it is not possible to define as yet the source of the Ir. This source can be the normal daily influx of extra-terrestrial matter, dissolved Ir in ocean water, Ir from deep mantle gasses or Ir from an exceptionally large extraterrestrial body that impacted at the end of the Maastrichtian. However, without doubt Ir was redistributed and concentrated during deposition and early diagenesis.


## Conclusion


The abnormally high iridium concentration at the K/T boundary of Stevns Klint can be explained as the result of abnormally low deposition rates and condensation-concentration during prolonged residence of the sediment in a favourable sub-oxic redox zone of bacterial metabolism, carbonate dissolution and clay mineral precipitation, at the basis of the zone of bioturbative mixing and/or storm reworking.

Abnormally low deposition rates were possibly the result of an exceptionally fast increase-decrease of the subsidence rate at the Maastrichtian-Danian boundary, that coincided with a precession period, characterized by a rather low storm intensity and associated low hydrodynamic energy.

The hypothesis that explains the presumed Late Cretaceous mass extinction and high Ir concentration at the K/T boundary by a catastrophic impact of a large extra-terrestrial Ir-rich body is not supported by the model for the dynamics of the depositional-early diagenetic environment that explains the genesis of the Stevns Klint sequence. The model shows how exceptionally favourable conditions for Ir precipitation occurred only locally and may have led to the local genesis of abnormally high Ir concentrations, despite a normal supply of Ir of unidentified source.

# 9 Self-organizing Model for Wavy Bedform Genesis

### Abstract

The Maastrichtian-Danian Tuffaceous Chalk of NW Europe is a subtropical, shallow marine, bioclastic carbonate silt- or sandstone. It forms tens of metres thick sequences, characterised by a regular succession of dm-m thick laterally continuous beds. Bedding is expressed by a rhythmic vertical variation of the grainsize, structures and early-diagenetic authigenic mineral concentrations. Tuffaceous Chalk cycles are fining-upwards cycles with a coarse-grained glauconitic, wavy laminated basis and with a fine-grained, homogeneously bioturbated, lithified, bored and encrusted top (hardground). Tuffaceous Chalk cycles presumably have been deposited during 20 ka precession periods and they reflect the cyclic variation of climate, storm frequency-intensity, hydrodynamic energy and deposition rates. The relation between the periodic variation of hydrodynamics, defining sediment erosion, transport and redeposition, and the simultaneous early-diagenetic lithification and decrease of erodability is investigated with a numerical model. The model is a self-organising parallel computer that allows the simulation of the genesis of regularly spaced bedforms (ripples, dunes, waves) from an initial randomly distributed bed height. A dynamic equilibrium is reached when a train of equi-dimensional bedforms moves down-current at constant velocity. When hydrodynamic energy, water depth and deposition rates are varied properly and when erodability is dependent on simultaneous lithification, then the model simulates the genesis of wavy bedded Tuffaceous Chalk cycles as have been observed in the field.

## Introduction

Tuffaceous Chalk is a well sorted bioclastic carbonate silt- or sandstone, deposited in shallow subtropical marine Maastrichtian-Danian seas. In NW Europe Tuffaceous Chalk crops out along the Gironde Estuary (SW France; Séronie Vivien, 1972), in quarries near Maastricht (SE Netherlands; Felder, 1975ab) and in a cliff section at Stevns Klint (SE Denmark; Surlyk, 1979).

The tens of metres thick deposits of sorted detrital carbonate clasts consist of dm-m thick fining-upwards cycles. The coarse-grained basis of the cycles is characterised by hummocky- and trough-cross stratification, indicating erosion, transport and deposition by (storm) waves. The fine-grained top of the cycles has been bioturbated and homogeneously mixed. Syn-sedimentary lithification of the carbonates is common and locally indicated by overgrown, bored and mineralised hardgrounds at the top of the cycles (Voigt, 1929, 1959, 1974; Bromley, 1968).

It is thought that the cycles have been deposited during, on average, 20 ka long precession periods (Cottle, 1989; Gale, 1989; Herrington et al., 1991; Chapter 5) and they reflect the repeated reworking of the seabottom during storms. Strength of reworking varied periodically, in phase with orbitally induced periodic variations of climate and oceanography and the simultaneous lithification of the carbonates influenced the depth of storm reworking and vice versa (Chapters 5, 6, 7). The effect of the feed-back between storm reworking and simultaneous lithification on bedding is a slow process and has received little attention. The more because the (Tuffaceous) Chalk sea, that was extensive, very shallow with a gently dipping coastal slope, occasionally characterised by relatively high hydrodynamic energy during storms and by rather low subsidence and deposition rates (cm-dm/ka; Håkansson et al., 1974), has no recent equivalents.

In order to further increase the understanding of the genesis of the wavy-bedded (Tuffaceous) Chalk cycles and to reconstruct the shallow marine sedimentary environment of NW Europe during the Late Cretaceous-Early Tertiary, numerical models are presented that allow the simulation of the genesis of sediment waves, while sediment simultaneously lithifies and discharge, water depth and deposition rates vary periodically. In order to test the model, the genesis of the wavy-bedded K/T boundary sequence of Stevns Klint (Denmark, Chapter 8) will be simulated.

## 9.1 The genesis of sediment waves

Grains at the surface of a bed, sheared by a gas or fluid current will, upon sufficient shear stress, be transported in the down-current direction. Where grains are removed from the bed, a depression appears (erosion) and where grains are added, the bed surface is elevated (deposition). At a current strength between a minimum value, where the grains are not moved and a maximum value, where all grains at the bed surface move, the morphology of the bed changes. Depressions and highs, several grain diameters wide and high, appear and are regularly distributed over the bed surface. The thus formed sediment waves grow until they reach a stable form and move down-current with a constant velocity (celerity) that is lower than the velocity of the fluid current.

The genesis of ripples, dunes or waves on a granular bed, sheared by a fluid current, has received considerable attention from a fundamental point of view. Numerous

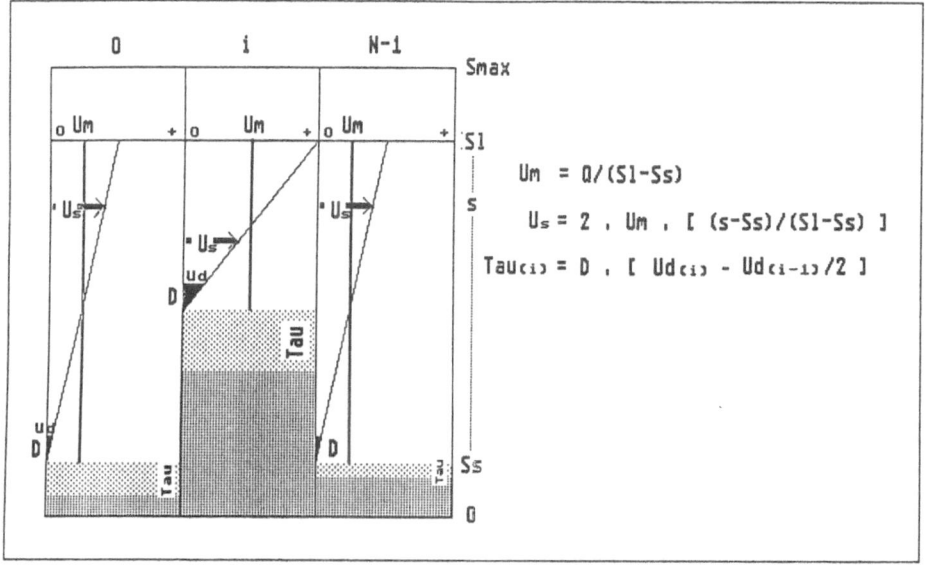

Figure 9.1 - Cellular channel of length N and depth Smax is closed (cell N-1 neighbours cell 0). Initial sediment surface (Ss) in the different cells is randomly chosen and lower than the water surface (Sl). Conditions in three successive cells i-1,i and i+1 are defined by the constant discharge (Q) that defines the depth (Sl-Ss) averaged current velocity Um. The current velocity Us at a layer s increases linearly from zero at the bottom to 2.Um at the water surface. The near-bottom velocity Ud at height D (grain-diameter) and the difference between the near-bottom velocities of cells i and i-1, define the bottom shear stress in cell i (Tau(i)). The amount of sediment that is eroded during dt, is proportional to tau(i) and is largest at up-current slopes and lowest at down-current slopes, despite equal depth averaged current velocities Um.

experiments have been performed that aimed to describe the relation between the various parameters that appear to control the bed-surface morphology, the migration rate of bedforms and the time required to establish an equilibrium morphology (cf. Allen, 1984; Baas, 1993). Important parameters are the grain diameter and density, the fluid viscosity and the distribution of the fluid velocity as a function of discharge, current depth and height above the bed surface.

Mathematical models have been presented that allow the calculation of the distribution of forces and the resulting redistribution of mass on a granular bed and in a shearing fluid current (Kennedy, 1964; Richards, 1980). A simple model is based on the notion that in a current, small bedforms move faster than larger bedforms and that in the course of time smaller bedforms will thus amalgamate into larger bedforms, in order to form more or less equally sized bedforms moving downcurrent in succession (Exner, 1931). According to such model, regularly spaced bedforms are generated on a bed with an initial random distribution of bed height, but the configuration is not stable and deteriorates into a single bedform eventually (Raudkivi & Witte, 1990).

Here a model is presented that also simulates the genesis of a ripple train from initially random bed-height distribution. In this model, dynamic equilibrium is reached when several more or less equally sized and spaced bedforms move downcurrent at constant velocity.

## 9.2 Cellular automaton for sediment wave genesis

The numerical model (Fig. 9.1) that simulates sediment transport in a fluid current is, in the broad sense, a cellular automaton (Wolfram, 1986; Anderson & Bunas, 1993). A cell space of N adjacent uniform cells is updated according to local rules that prescribe the change of properties of a cell, from the moments t to t+dt, as a function of the properties of this cell and its neighbouring cells at the moment t. Initially (t=0), the water depth (sediment height) is defined (Procedure 1), followed (during t=0 to t=T) by procedures for fluid velocity, erosion, transport and deposition (Procedures 2, 3, 4 and 5) that are repeated each step dt.

```
+ = addition  - = substraction  * = multiplication  / = division
pi = 3.14..
sin( ) = goniometric function
A(m,n) = array of m x n sites

FOR t=0 TO t=T
    IF t=0
            GO TO PROCEDURE 1 (initial depth)
    ELSE
            GO TO PROCEDURE 2 (fluid velocity)
            GO TO PROCEDURE 3 (erosion)
            GO TO PROCEDURE 4 (transport)
            GO TO PROCEDURE 5 (deposition)
    ENDIF
NEXT t
```

PROCEDURE 1 - (initial depth) A channel with maximum depth Smax and length N is simulated by an array of N cells. The channel is a closed conduit and cell N-1 is a neighbour of cell 0. Initially (t=0), the water level and the randomly chosen amount of sediment are defined.

```
FOR i=0 TO i=N-1
        Sl(t) = Sl0
        Ss(i,t) = Ssmin+random(Ssmax-Ssmin)
NEXT i

Sl(t) = water level during t
Sl0 = water level during t=0 (0<Sl0<Smax)
```

Smax = maximum channel depth
Ss(i,t) = sediment surface at cell i during t
random(x) = random integer number between 0 and x
Ssmin = minimum height of initial sediment layer
Ssmax = maximum height of initial sediment layer
0 < Ssmin,Ssmax < Smax
RETURN

PROCEDURE 2 (Fluid velocity) - The discharge (Q) or the volume of fluid that streams through a cell per unit time (dt) is equal for each cell. The average current velocity in a cell is thus equal to the discharge divided by the water depth. For the sake of simplicity, the depth dependent current velocity (U(i,s)) is zero at the sediment surface, equal to the average current velocity at half the water depth and twice the average current velocity at the water surface.

```
FOR i=0 TO i=N-1
        FOR s=Ss(i,t) to s=Sl(t)
                Um = Q(t)/(sl(t)-Ss(i,t))
                U(i,s) = 2*um*((s-Ss(i,t))/(Sl(t)-Ss(i,t)))
        NEXT s
NEXT i
```

Um = average current velocity in cell i
Q(t) = discharge during t+dt in all cells i
Sl(t)-Ss(i,t) = water depth in cell i
U(i,s) = current velocity at layer s above the sediment surface in cell i
RETURN

PROCEDURE 3 (Erosion) - The grains start to move at a certain minimum shear stress. The stress is proportional to the near-bottom current velocity and the sediment grain diameter. It appeared empirically that the shear stress has to be defined as a function of the near-bottom current velocity and the difference between near bottom current velocities of adjacent cells (spatial acceleration). The amount of grains that goes into motion (erosion) is proportional to the shear stress.

```
FOR i=0 TO i=N-1
        IF i=0
                i-1=N-1 (closed loop)
        ENDIF
        U_tot=0
        dU_tot=0
        FOR s=Ss(i,t) TO s=Ss(i,t)+D
                U_tot = U_tot+U(i,s)
                dU_tot = dU_tot+(U(i,s)-U(i-1,s))
        NEXT s
        Tau(i) = U_tot+dU_tot
        IF Tau(i) < Tau_min
                Tau(i) = Tau_min
        ENDIF
        IF Tau(i) > Tau_max
                Tau(i) = Tau_max
        ENDIF
        Ss(i,t+1) = Ss(i,t)-Tau(i) (erosion)
        C(i) = C(i)+Tau(i)
NEXT i
```

U_tot = summed current velocity at cell i from the sediment surface to D.

dU_tot = difference of summed current velocities U_tot at cell i and i-1

D = grain diameter

C(i) = amount of grains in motion at cell i

Tau(i) = bottom shear stress at cell i

Tau_min = minimum shear stress for grain motion (Tau_min > 0)

Tau_max = maximum shear stress (Tau_max <<Smax)

note that, in order to account for gravity forces, water depth differences and thus the bottom shear stress defining velocity differences, can be limited by an "avalanche" procedure, so that Tau_max is limited and becomes a function of the angle of initial yield.

RETURN

PROCEDURE 4 (Transport) - Grains that have been eroded are transported from cell i-1 to cell i according to a diffusion rule at a rate that depends on the amount of sediment in motion (C(i)).

```
FOR i=0 TO i=N-1
        IF i=0
                        i-1=N-1 (closed loop)
        ENDIF
        dC(i) = (1-dc)*C(i)+dc*C(i-1)
NEXT i
FOR i=0 TO i=N-1
        C(i) = dC(i)
NEXT i
```

dc = diffusion coefficient 0.5<dc<1

dC(i) = local array

RETURN

PROCEDURE 5 (Deposition) - Part of the grains is deposited after transport

```
FOR i=0 TO i=N-1
        Ss(i,t) = Ss(i,t)+dp*C(i)
        C(i) = C(i)-dp*C(i)
NEXT i
```

dp = deposition coefficient 0.5<dp<1

RETURN

*Opposite page:*

Figure 9.2 - (upper) Initial sediment surface in a channel of 80 cells and a randomly chosen depth.

Figure 9.3 - (middle) Sediment surface after 640 iterations and the remains of previous sediment surfaces (current from left to right).

Figure 9.4 - (lower) Sediment surface after 1920 iterations. Near-equilibrium condition has been reached.

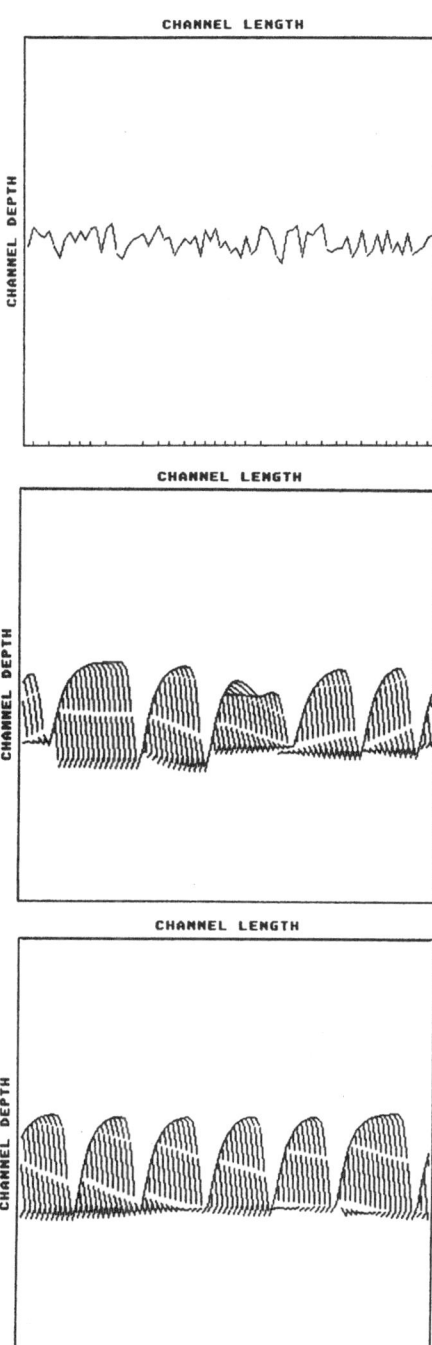

## 9.2.1 Simulation results

After the initial conditions have been set at t=0, velocity, erosion, transport and deposition procedures have been repeated T times. The initial random distribution of depth changes into a new distribution that represents a dynamic equilibrium characterised by a number of asymmetric sediment waves with constant wave-length and wave-height and moving at constant velocity down-current (Figs. 9.2-9.8). The concentration of sediment in motion increases from trough to crest and decreases from crest to trough in the direction of the current (not depicted).

The depth-length model can be easily extended to a depth, length and width model, simulating the genesis of 3D waves (Fig. 9.9). In that case, bottom shear stress of cell i,j is a function of the current velocity U(i,j,s) and of the average spatial acceleration (procedure 3):

$$Tau(i,j) = 2*U(i,j,s) - (U(i-1,j-1,s) + U(i-1,j,s) + U(i-1,j+1,s))/3$$

The sediment transport in the 3D model is defined by the rule (procedure 4):

$$dC(i,j) = (1-(dc1+2*dc2))*C(i,j) + dc1*C(i-1,j) + dc2*C(i-1,j-1) + dc2*C(i-1,j+1)$$
$$dc1 > dc2 \text{ and } 0.5 < dc1+2*dc2 < 1$$

The model results shown above, are comparable to the patterns produced during the genesis of bedforms of different dimensions, such as ripples, megaripples and waves. One notes furthermore that the discrete succession of erosion, transport and deposition procedures, allows the simulation of storm successions with oscillatory flow (erosion), superimposed current (diffusive transport) and storm waning (deposition), and of the genesis of wavy storm erosion surfaces.

## 9.2.2 Variation of discharge, depth, and deposition

During the genesis of an about 20 ka precession cycle, the average storm intensity, hydrodynamic energy and water depth increased and decreased again. It is assumed that, when sea level did not fluctuate and average hydrodynamic energy remained constant, a dynamic equilibrium existed during which deposition rates were equal to the subsidence rate. However, when the hydrodynamic energy increased, depth increased and deposition rates decreased and became lower than the subsidence rate, while during decrease of the hydrodynamic energy, depth decreased and deposition rates increased and exceeded the subsidence rate.

During the simulation period t=0 to t=T during which S layers of sediment are deposited,

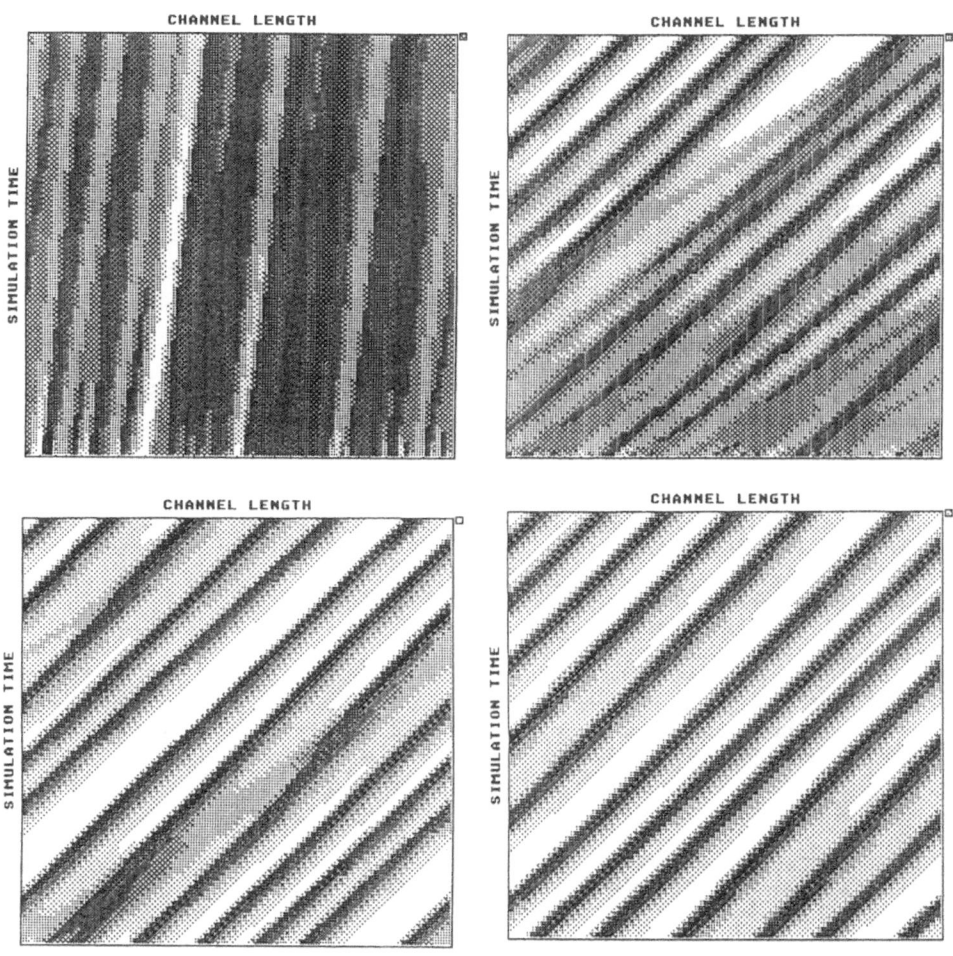

Figure 9.5 - (upper left) Evolution of bed surface during 80 iterations (t=0 to t=79) from an initial random distribution. Shades of grey define the relative bed-height or water depth (dark=deep and light=shallow). Current from left to right. Time from basis to top. Note the amalgamation of rapidly moving small bedforms with slowly moving larger bedforms and the decrease of the number of bedforms in time.

Figure 9.6 - (upper right) As Fig. 9.5, but for 640 iterations (t=0 to t=639).

Figure 9.7 - (lower left) Continuation of Fig. 9.6 (t=640 to t=1279)

Figure 9.8 - (lower right) Continuation of Fig. 9.7. Near-equilibrium conditions are reached during t=1280 to t=1919.

subsidence rate is presumed to be constant. Sea level rises relatively to a subsiding datum plane Sl0 at a constant rate (procedure 6). The storm intensity (procedure 7), water depth (procedure 8) and the discharge (procedure 9) are defined as a function of the cyclic climate variation during a 20 ka precession period.

```
FOR t=0 TO t=T
   GO TO PROCEDURE 6 (sea level)
   GO TO PROCEDURE 7 (storm intensity)
   GO TO PROCEDURE 8 (water depth)
   GO TO PROCEDURE 9 (discharge)
NEXT t

PROCEDURE 6 (Sea level)
   Sl(t) = Sl0+t*(Smax-Sl0)/T
   Sl(t) = sea level at t
   Sl0 = sea level at t=0
   Smax = maximum sea level
RETURN
```

PROCEDURE 7 (Storm intensity) - The storm intensity and hydrodynamic energy vary periodically during T. The variation is defined by a theoretic function with periods of 20 and 100 ka and an amplitude that varies like the precession index.

```
E(t)=(0.5+0.5*(A20+(A100*(0.5+0.5*sin((2*pi/100)*t))))*sin((2*pi/20)*t))
E(t) = periodically varying hydrodynamic energy 0<E(t)<1
RETURN
```

PROCEDURE 8 (Water depth) - The water depth (depth of the storm erosion surface below sea level) varies periodically and is proportional to the hydrodynamic energy. The position of the storm erosion surface is defined for the period T

```
Ss(t)=Sl(t)-DES*E(t)

Ss(t) = storm erosion surface at t
DES = maximum depth of storm erosion surface when E(t)=1
RETURN
```

PROCEDURE 9 (Discharge) - The average current velocity (discharge) is proportional to the hydrodynamic energy

```
Q(t)=Qmax*E(t)
Q(t) = discharge at t
Qmax = maximum discharge
RETURN
```

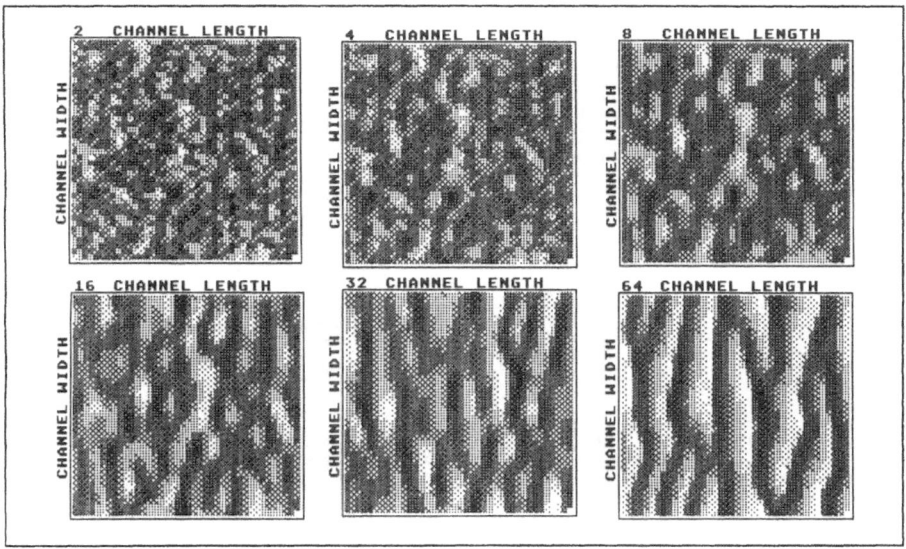

Figure 9.9 - Wave genesis in a 3D space of 40 x 40 cells (light=shallow, dark=deep). Current from left to right. Genesis of downcurrent migrating sediment waves from an initial random distribution in 64 iterations, shown at t = 2, 4, 8, 16, 32 and 64.

### 9.2.3 Simultaneous lithification

As water depth and discharge have been defined, the genesis of wavy bedding during T can be simulated, using the numerical model for the genesis of bedforms (procedures 1-5), while sea level, storm intensity, water depth and discharge change (procedures 6-9) and sediment is added or removed according to the defined change of hydrodynamic energy (procedure 10). At the same time sediment is cemented in a redox zone below the storm erosion surface (procedure 11) and consequently the erodability decreases (procedure 12) (Figs. 9.10, 9.11).

```
FOR t=0 TO t=T
  IF t=0
        GO TO PROCEDURE 1 (initial depth)
  ELSE
        GO TO PROCEDURE 2 (fluid velocity)
        GO TO PROCEDURE 3 (erosion)
        GO TO PROCEDURE 4 (transport)
        GO TO PROCEDURE 5 (deposition)
        GO TO PROCEDURE 10 (deposition/erosion)
        GO TO PROCEDURE 11 (cementation)
        GO TO PROCEDURE 12 (decrease of erodability)
  ENDIF
NEXT t
```

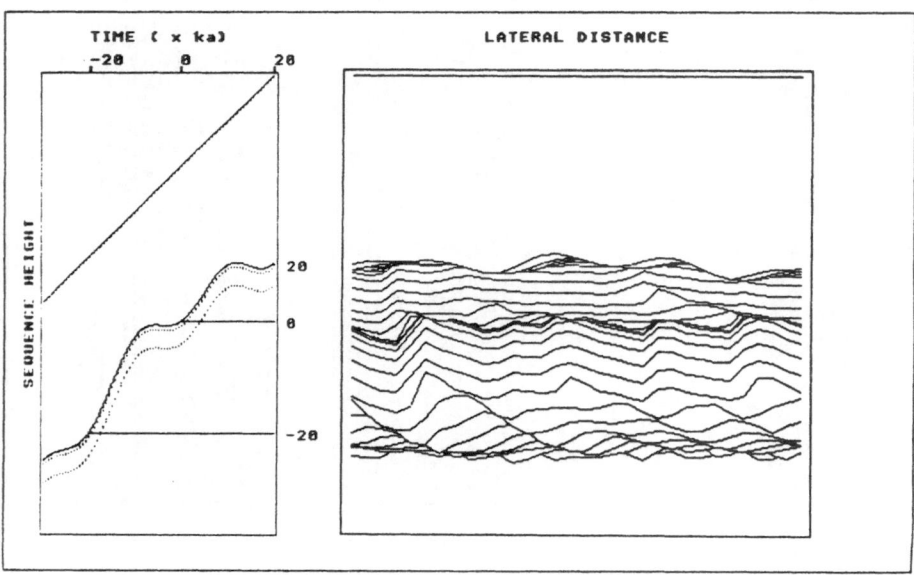

Figure 9.10 - The result of the simulation of the genesis of the last Maastrichtian and first Danian precession cycles exposed at Stevns Klint (Denmark) (Chapter 8, Fig. 8.2). Left box: during a 50 ka period, sea level and subsidence rate remain constant while water depth and depth of the storm erosion surface increase and decrease in phase with climatically induced increase-decrease of the storm frequency and intensity. While the hydrodynamic energy increases, deposition rates decrease. The discharge and wavy storm erosion surface celerity increase, while wave-height decreases. During hydrodynamic energy decrease, the reverse occurs.

PROCEDURE 10 (deposition/erosion) - The average depth of the storm erosion surface (water depth) is adapted to the calculated water depth as a function of the sea-level rise and periodically varying storm intensity, by means of the change of the concentration of sediment in motion.

```
Ss_tot=0
FOR i=0 TO i=N-1
        Ss_tot = Ss_tot+Ss(i,t)
NEXT i
FOR i=0 TO i=N-1
        C(i) = C(i)+(Ss(t)-(Ss_tot/N))
NEXT i
```

Ss_tot/N = average of the storm erosion surface Ss(i,t) in cells 0 to N-1 during t

Ss(t)-Ss_tot/N = difference between the actually present average storm erosion surface and the storm erosion surface Ss(t) defined above as a function of the hydrodynamic energy E(t) (procedure 8)

C(i) = concentration of sediment in motion at cell i is increased or decreased in order to provoke deposition or erosion and to correct for the difference between predicted storm erosion surface (Ss(t)) and observed average storm erosion surface (Ss_tot/N)

RETURN

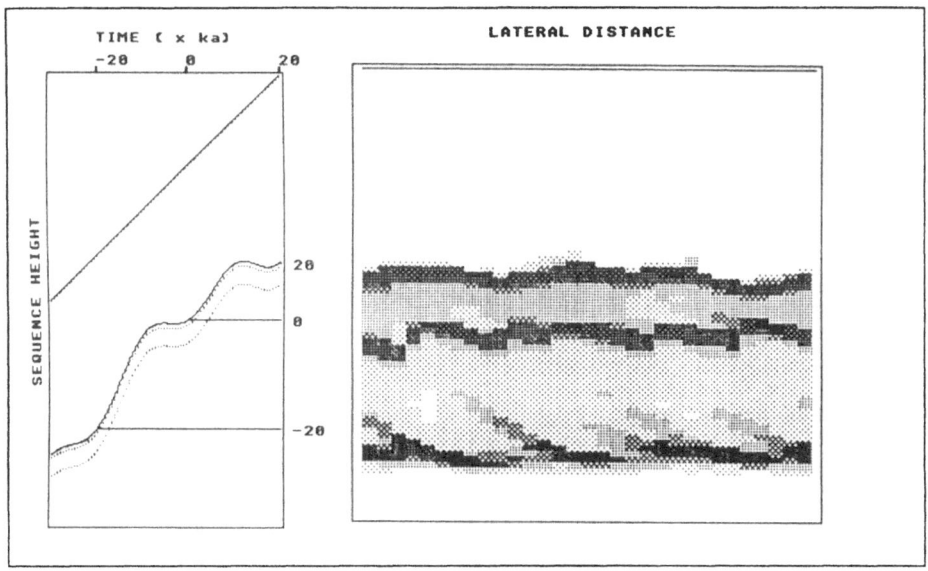

Figure 9.11 - As Fig. 9.10, but with depiction of the authigenic silica/cement concentrations.

PROCEDURE 11 (cementation) - The cementation of the sediment in cell i is, at each step dt, increased in a redox zone with constant reaction rate, constant thickness and at constant depth below the storm sediment surface of cells 0 to N-1.

```
FOR i=0 TO i=N-1
        FOR s=Ss(i,t)-RZDmin TO s=Ss(i,t)-RDZmax
                Cem(i,s)=Cem(i,s)+R
        NEXT s
NEXT i
```

RDZmin,RDZmax = minimum, maximum depth of redox zone below storm erosion surface at cell i.
Cem(i,s) = degree of cementation in layer s at cell i.
R = reaction rate.
RETURN

PROCEDURE 12 (decrease of erodability) - The erosion of the sediment is a function of bottom shear stress and the degree of cementation. The bottom shear stress is reduced when the degree of cementation exceeds a minimum value. Cement is destroyed (Cem(i,s) set zero) if cemented layers are eroded.

```
FOR i=0 TO i=N-1
        Cem_tot=0
        FOR s=Ss(i,t) TO s=Ss(i,t)-Tau(i)
                Cem_tot=Cem_tot+Cem(i,s)
        NEXT s
```

```
        IF Cem_tot>Cemmin
                Tau(i)=Tau(i)*(Cemmin/Cem_tot)
        ENDIF
NEXT i
```

Cem_tot = degree of lithification in cell i, summed over s layers below the storm erosion surface to the erosion depth that is proportional to the shear stress Tau(i).

Cemmin = minimum degree of cementation at which erosion still occurs to a depth that is proportional to the shear stress only.

RETURN


## Discussion


The dimensionless self-organizing parallel computer that has been developed in order to simulate the genesis of wavy bedding in (Tuffaceous) Chalk sequences appears to behave like natural dynamic processes and produces patterns of sediment distribution that can be observed in the field. The model simulates the genesis of wavy storm erosion surfaces when discharge and water depth vary according to periodic changes in the storm intensity and the deposition rates, while sediment simultaneously lithifies and erodability decreases. The storms are simulated by a succession of erosion, transport and deposition procedures that is repeated during the simulation period (T) and that represent the erosion by wave and the transport by a superimposed uni-directional current. Of course, like every model, also this model is only a representation of the understanding of the processes of interest and not of the processes itself. The value of the model lies in the fact that it is a universal logic language which is intended to stimulate the discussion of complex dynamic processes in an accurate and condensed form.

The model contributes to a better understanding of the facies distribution in (Tuffaceous) Chalk and to the reconstruction of the Cretaceous-Tertiary marine environment of NW Europe. Unfortunately, the model lacks a proper definition of the relation between the variations of hydrodynamic energy, depth and discharge. This relation is not trivial, as has been concluded from the preliminary results of simulation models that aim at the calculation of the momentum distribution in a shallow marine basin of variable depth, variable sea bottom dip and with energy introduction by oceanic waves and/or by wind surface shear.

# Conclusion

Sediment erosion by an oscillating, shearing fluid flow with superimposed current, over a granular bed-surface, initially with a random depth distribution, leads ultimately to the genesis of wavy erosion surfaces with regularly spaced crests and troughs, that are of near-equal form and that move in the downcurrent direction. The presented numerical model simulates the evolution of a bed-surface with initial random height distribution towards a stable wave train, moving downcurrent.

Subtropical marine fine-grained detrital carbonates like the Maastrichtian-Danian Tuffaceous Chalk of NW Europe have been deposited at rates of cm-dm/ka. While bedforms migrated, sediment lithified at some depth below the sediment surface.

The above numerical model that simulates the genesis of wavy bedforms, while sediment simultaneously lithifies and while hydrodynamic energy, water depth and deposition rates vary periodically, produces simulated sequences with a distribution of depositional structures and authigenic mineral concentrations, fairly similar to those observed in the field.

Thus (Tuffaceous) Chalk cycles can not only be characterised by their lithofacies, but also by the dynamics that led to their genesis so that eventually the dynamics of the Cretaceous-Tertiary marine environment of NW Europe can be reconstructed.

# Epilogue

Nineteenth century geologists knew the paleontology of the Chalk and had proposed a bio-/lithostratigraphic subdivision of the Late Cretaceous/Early Tertiary, shallow (sub)tropical marine carbonate mudstones. During this century the stratigraphy of the Chalk was refined. Recently developed tools have allowed large-scale subsurface exploration of bedding, as well as micro-scale investigations of the smallest fossils and minerals in Chalk.

Despite the much increased knowledge of the facies distribution, hardly anything is known about the environment in which Chalk formed and about the dynamics of the depositional and early diagenetic processes that have defined the characteristic bedding in Chalk.

Because Chalk is poorly exposed relative to its extensive occurrence and because Chalk deposits have been displaced during Tertiary lateral fault-block migration, it is necessary to be able to reconstruct the dynamics of the marine environment from the sedimentary characteristics of small exposures, so that the large-scale spatial/temporal variation of the environmental dynamics and of the spatial distribution of the bio-/lithofacies of Chalk can be predicted more accurately.

It has been shown how, with the help of computers and numerical models, the genesis of different Chalk sequences can be simulated, while several model parameters are varied. Thus the Chalk sequences cannot only be characterised by their paleontology and lithology, but also by the parameters that characterised the dynamics of the depositional and the early diagenetic environment.

Future research might focus on the further development of the proposed models, so that the genesis of the large-scale, basin-wide distribution of Milankovitch bedding in Chalk, as for example reflected by gamma-ray (glauconite) profiles of boreholes and by acoustic impedance variations (lithification) of seismic profiles, can be simulated. The extended models may assist in further reconstructing the dynamics of NW Europe during the Late Cretaceous-Early Tertiary and in predicting the occurrence of Chalk deposits of particular importance.

# References

* cited in Binkhorst van den Binkhorst (1859)

Abildgaard, S. (1759) *Beskrivelse over Stevns Klint og dens naturlige Maerkvaerdigheder.* 50 pp. Kobenhavn.

Aigner, T. (1979) Coquinal tempestites in the upper Muschelkalk, Triassic, southern West Germany. *Neues Jahrb. Geol. Paleontol. Abh.,* **157**, 326-343.

Aigner, T. (1982) Calcareous tempestites: Storm dominated stratification in Upper Muschelkalk limestones (Middle Trias, SW Germany) In: *Cyclic and Event Stratification* (Ed. by Einsele G., Ricken W., & Seilacher A.) pp. 100-198. New York, Springer Verlag. 955 pp.

Aigner, T. (1985) *Storm depositional systems: dynamic stratigraphy in modern and ancient shallow-marine sequences.* (Lect. Notes Earth Sci. 3), Springer, Berlin Heidelberg New York, 174 pp.

Albers, H.J. (1976) Feinstratigraphie, Faziesanalyse und Zyclen des Untercampans (Vaal-sergruensand = Hervien) von Aachen und dem niederländisch-belgischen Limburg. *Geol. Jb.,* **A 34**, 3-68.

Albers, H.J. & Felder, W.M. (1979) Feuersteine als Indikatoren der Quantifizierung und Datierung der Karbonatlösung am Nordwest-Rand des Rheinischen Schildes. *Neder-landse Geol. vereninging,* **Staringia 6**, 18-22.

Allen, J.R.L. (1984) *Sedimentary structures, their character and physical basis.* Developments in Sedimentology, **30**, Elsevier, Amsterdam. 593 + 663 pp.

Alvarez, L.W., Alvarez, W., Asaro, F. & Michel, H.V. (1980) Extraterrestrial cause for the Cretaceous-Tertiary extinction. *Science,* **208**, 1095-1108.

Anderson, R.S. & Bunas, K.L. (1993) Grain size segregation and stratigraphy in aeolian ripples modelled with a cellular automaton. *Nature,* **365**, 740-743.

Applin, K.R. (1987) The diffusion of dissolved silica in dilute aqueous solutions. *Geochim. Cosmochim. Acta.,* **51**, 2147-2151.

Arthur, M.A. & Dean, W.E. (1991) A Holistic Geochemical Approach to Cyclomania: Examples from Cretaceous Pelagic Limestone Sequences. In: *Cycles and events in stratigraphy* (Ed. by Einsele, Ricken, Seilacher) pp. 126-167. Springer-Verlag Berlin-Heidelberg-New York, 955 pp.

Baas, J.H. (1993) Dimensional analysis of current ripples in recent and ancient depositional environments. *Geologica Ultraiectina*, **106**, 199 pp.

Baes, Jr. C.F. & Mesmer, R.E. (1976) *The hydrolysis of cations.* Wiley, New York. 489 pp.

Barron, E.J., Harrison, C.G.A., Sloan II, J.L. & Hay, W.W. (1981) Paleogeography, 180 million years ago to the present. *Eclogae geol Helv.*, **74/2**, 443-470.

Bathurst, R.G.C. (1971) Carbonate sediments and their diagenesis. *Developments in Sedimentology*, **12**, Elsevier Scientific Publishing Company. 658 pp.

Batten, D.J. Dupagne-Kievits, J. & Lister, J.K. (1988) Palynology of the Upper Cretaceous Aachen Formation of Northeast Belgium. In: *The Chalk District of the Euregio-Rhine* (ed. by Streel, M. & Bless, M.J.M.), pp. 95-103. Nat. hist. Mus. Maastr. & Lab. Pal. Univ. d'Etat Liège. 117 pp.

Bayle, E. (1857/1858) Sur les Rudistes découverts dans la craie de Maastricht. *Bull. Soc. geol. France*, **2**, (15), 210-218.

Berner, R.A. (1969) Migration of iron and sulphur within anaerobic sediments during early diagenesis. *Am. Jour. Sci.*, **267**, 19-42.

Berner, R.A. (1970) Sedimentary pyrite formation. *Am. Jour. Sci.*, **268**, 13-23.

Berner, R.A. (1971) *Principles of Chemical Sedimentology.* McGraw-Hill, New York, 240 pp.

Berner, R.A. (1980) *Early Diagenesis.* Princeton series in geochemistry, 241 pp.

Berger, A.L. (1979) Long-Term Variations of Daily Insolation and Quaternary Climatic Changes. *Journal of the atmospheric sciences*, **35**, 2362-2367.

Berger, A.L. (1988) Milankovitch theory and climate. *Rev. Geophys.*, **26**, 624-657.

Binkhorst van den Binkhorst, J.T. (1857) Neue Krebse aus der Maestrichter Tuffkreide. *Verh. naturhist. Ver. preuss. Rheinl. Westf.*, **XIV** (3), 107-110.

Binkhorst van den Binkhorst, J.T. (1859) *Esquisse Géologique et Paléonthologique des couches crétacées du limbourg, et plus specialement de la craie tuffeau.* Maastricht. 268 pp.

Binkhorst van den Binkhorst, J.T. (1861) Monographie des gastropodes et des céphalopodes de la craie supérieur du Limbourg, suivie d'une description de quelques espéces de crustacés du méme dépot crétacé. 83 pp. + 44 pp.

Birdsall, C.B. & Steward, R.G. (1978) Hurricane impact on the continental shelf stratigraphy, Gulf of Mexico. *Geol. Soc. America Abs.*, **10**, 162.

Bischoff, J.L. & Sayles, F.L. (1972) Pore fluid and mineralogical studies of recent marine sediments: Bauer Depression of East Pacific Rise. *J. Sedim. Petrol.*, **41**, 711-724.

Black, M. (1953) The constitution of the Chalk. *Proc. geol. Soc. Lond.*, **1499**, 81-86.

Bless, M.J.M (1988) Upper Campanian lithofacies and ostracode assemblages in South Limburg and NE Belgium. In: *The Chalk District of the Euregio-Rhine* (ed. by Streel, M. & Bless, M.J.M.), pp. 57-68. Nat. hist. Mus. Maastr. & Lab. Pal. Univ. d'Ètat Liège. 117 pp.

Bless, M.J.M. & Robaszynski, F. (1988) Microfossils from the Vaals greensand (lower Campanian) in NE Belgium and South Limburg and from the Ems greensand (Santonian) in SW Münsterland (FRG). In: *The Chalk District of the Euregio-Rhine* (ed. by Streel, M. & Bless, M.J.M.), pp. 85-93. Nat. hist. Mus. Maastr. & Lab. Pal. Univ. d'Etat Liège. 117 pp.

Bond, G. (1978) Speculations on real sea-level changes and vertical motions of continents at selected times in the Cretaceous and Tertiary periods. *Geology*, **6**, 247-250.

Bosquet, J. (1847) Description des Entomostracés fossiles de la craie de Maestricht. *Mém. r. Soc. Sci. Liége*, **IV**.

Bosquet, J. (1854) Les crustacés fossiles du terrain crétacé du Limbourg. *Verh. Comm. geol. Beschr. Kaart Nederland*, **II**, 11-183.

Bosquet, J. (1859) Monographie des Brachiopodes fossiles du terrain crétacé supérieur du Duché de Limbourg. partie: Craniadae et Terebratulidae. *Mém. Serv. Descr. geol. de la Néerl*, **3**, 1-150.

Bottjer, D.J. & Ausich, W.I. (1986) Phanerozoic development of tiering soft substrata suspension feeding communities. *Paleobiology*, **12**, 400-420.

Breddin, H. (1929) Die Bruchfaltentektonik des Kreidegebirges im nordwestlichen Teil des Rheinisch-Westfälischen Steinkohlenbeckens. *Glückauf*, **65**, 1157-1168/1193-1198.

Breddin, H. (1932) Über die tiefsten Schichten der Aachener Kreide sowie eine senone Einebnungsfläche und Verwitterungsrinde am Nordwestabfall des Hohen Venns. *Centralbl. Min. Geol. Pal.*, **B**, 593-613.

Breddin, H., Bruehl, K.H. & Dieler, H. (1963) Das Blatt Aachen-Nordwest der praktisch-geologischen Grundkarte 1 : 5000 des Aachener Stadtgebietes. *Geol. Mitt.*, **1**, 251-428.

Bromley, R.G. (1965) *Studies in the Lithology and Conditions of Sedimentation of the Chalk Rock and Comparable Horizons*. Thesis, Univ. London, London, 335 pp, unpublished.

Bromley, R.G. (1968) Burrows and borings in hardgrounds. *Meddr. Dansk Geol. Foren, Kobenhavn*, **18**, 247-250.

Bromley, R.G. (1975) Trace fossils at omission syrfaces. In: *The study of Trace Fossils* (ed. by Frey M.D.). Springer-Verlag New York Inc. pp. 399-428.

Bromley, R.G. (1979) Chalk and Bryozoan limestone: Facies, Sediments and Depositional Environments. In: *Cretaceous-Tertiary boundary events. Symposium. I, The Maastrichtian and Danian of Denmark*. (Ed. by Birkelund, T. & Bromley, R.G.) pp. 16-32. University of Copenhagen. 210 pp.

Bromley, R.G. & Gale, A.S. (1982) The lithostratigraphy of the English Chalk Rock. *Cretaceous Research*, **3**, 273-306.

Bromley, R.G. & Ekdale, A.A. (1986) Flint and fabric in the European chalk. In: *The Scientific Study of Flint and Chert* (ed. by Sieveking, G. & Hart, M.B.), pp. 71-82. Cambridge University Press. 290 pp.

Bromley, R.G. Schulz, M.G. & Peake, N.H. (1975) Paramoudras: Giant flints, long burrows and the early diagenesis of chalks. *K. Danske. Vidensk. Selskab biol. Skr.*, **20, 10**, 31 pp.

Buchardt, B. & Jørgensen, N.O. (1979) Stable isotope variations at the Cretaceous/Tertiary boundary in Denmark. *Cretaceous-Tertiary boundary events. Symposium. I, The Maastrichtian and Danian of Denmark*. (Ed. by Birkelund, T. & Bromley, R.G.) pp. 16-32. University of Copenhagen. 210 pp.

Buckland, (1839)*

Burst, F.J. (1958) "Glauconite" pellets: their mineral nature and applications to stratigraphic interpretations. *Bull. Amer. Assoc. Petrol. Geol.*, **39**, 484-492.

Buurman, P. & van der Plas, L. (1971) The genesis of Belgian and Dutch flints. *Geol. Mijnbouw*, **50**, 9-28.

Camper, P. (1786) Philosoph. Trans. L. XXVI.*

Calembert, L. (1953) Sur l'existence regionale d'une hard ground et d'une lacune stratigraphique dans le Crétacé supérieur du Nord-Est de la Belgique. *Bull. Cl. Sci. Acad. roy. Belg.*, **5/39**, 724-733.

Calembert, L. (1956) Sur l'extension d'une lacune stratigraphique dans le Crétacé supérieur du Pays de Herve et du Limbourg hollandais. *Ann. Soc. Geol. Belg.*, **79**, 413-423.

Calvert, S.E. (1974) Deposition and diagenesis of silica in marine sediments. In: *Pelagic Sediments: on Land and under the Sea.*(Ed. by Hsü, K.J & Jenkyns, H.C.). *Int. Ass. Sediment Spec. Publ.*,**1**,273-299.

Christensen, W.K. (1988) Upper Cretaceous Belemnites of Europe: State of the Art. In: *The Chalk District of the Euregio-Rhine* (ed. by Streel, M. & Bless, M.J.M.), pp. 5-16. Nat. hist. Mus. Maastr. & Lab. Pal. Univ. d'État Liège. 117 pp.

Christensen, L., Fregerslev, S., Simonsen, A. & Thiede, J. (1973) Sedimentology and depositional environment of Lower Danian Fish Clay from Stevns Klint, Denmark. *Bull. geol. Soc. Denmark*, **2**, 192-212.

Clayton, C.J. (1986) The chemical environment of flint formation in Upper Cretaceous chalks. In: *The Scientific Study of Flint and Chert* (Ed. by Sieveking, G. & Hart, M.B.), pp. 43-54. Cambridge University Press. 290 pp.

Colodner, D.C., Boyle, E.A., Edmond, J.M. & Thomson, J. (1992) Post-depositional mobility of platinum, iridium and rhenium in marine sediments. *Nature*, **358**, 402-404.

Cooper, M.R. (1977) Eustasy during the Cretaceous: Its implications and importance. *Paleogeography, Paleoclimatology, Paleoecology*, **22**, 1-60.

Coquand, H. (1857) Position des Ostrea columba et biauriculata dans le group de la craie inférieur. *Bull Soc. Géol. de la France 2e série*, **14**, 749 pp.

Cottle, R.A. (1989) Orbitally mediated cycles from the Turonian of Southern England: Their potential for high-resolution stratigraphic correlation. *Terra Nova*, **1**, 426-431.

Davreux, C.J. (1833) Essai sur la constitution géognostique de la province de Liège. *Mem. cour. Acad. Bruxelles*, 9-4.*

Debey, M.H. (1857) Tageblatt der 33 Versammlung Deutscher Naturforscher und Arzte in Bonn. **6**.*

Debey, M.H. (1849) Entwurf zu einer Geognostisch-geogenetischen Darstellung der gegend von Aachen. *Verh. dt. Naturforsch. u. Aerzte, geol.-miner. Sekt.*

De Boer, P.L. (1983) Aspects of Middle Cretaceous Pelagic sedimentation in southern Europe. Production and storage of organic matter, stable isotopes and astronomical influences. *Geologica Ultraiectina*, **31**, 111 pp.

De Boer, P.L. (1991a) Astronomical cycles reflected in sediments. *Zentrl Abl. für Geologie und Palaeontologie*, **8**, 911 - 930.

De Boer, P.L. (1991b) Pelagic Black Shale-Carbonate Rhythms: Orbital Forcing and Oceanographic Response. In: *Cycles and events in stratigraphy* (Ed. by Einsele, Ricken, Seilacher) pp. 63-79. Springer-Verlag Berlin-Heidelberg-New York, 955 pp.

Deroo, G. (1966) Cytheracea (Ostracodes) du Maastrichtien de Maastricht (Pays-Bas) et des régions voisines; résultats stratigraphiques et paléonthologiques de leur étude.

*Med. Geol. Stichting*, **C46**, 1-197.

Desor, E. (1846) Sur le terrain danien, nouvel étage de la craie. *Bull. Soc. geol. Fr.*, **4**, 179-182.

D'Orbigny, A.C.V.D. (1840-1842) *Paléontologie francaise; terrains crétacés I*, Masson, Paris.

Dott, R.H. & Bourgois, J. (1982) Hummocky stratification: significance of its variable bedding sequences. *Geol. Soc. Am. Bull.*, **93**, 603-680.

Drever, J.L. (1982) *The geochemistry of natural waters*. Prentice-Hall, Englewood Cliffs, N.J., 338 pp.

Drobner, E., Huber, H., Wächterhäuser, G., Rose, D. & Stetter, K. O. (1990) Pyrite formation linked with hydrogen evolution under anaerobic conditions. *Nature*, **346**, 742-744.

Dumont, A.H. (1832) Mémoire sur la constitution géologique de la province de Liège.[*]

Dumont, A.H. (1847) Mémoire sur les terrains ardennais et rhenan de l'Ardenne, du Rhin, du Brabant et du Condroz. Première partie (terrain ardennais). *Mémoires couronnés de l'Académie royale de Belgique*, **20**, 163 pp.[*]

Dumont, A.H. (1849) Rapport sur la carte géologique du Royaume, *Bull. Acad. Roy. Sci. Lettres Beaux-Arts Belgique*, **16 (11)**, 351-373.[*]

Einsele, G. Ricken, W. & Seilacher, A. (1991) Cycles and Events in Stratigraphy - Basic Concepts and Terms. In: *Cycles and events in stratigraphy* (Ed. by Einsele, Ricken, Seilacher) pp. 1-19. Springer-Verlag Berlin-Heidelberg-New York, 955 pp.

Ehrenberg, (1812) Das unsichtbar wirkende organische Leben, Vorlesung zu Berlin, am 12 Febr.[*]

Ekdale, A.A. & Bromley, R.G. (1991) Analysis of Composite Ichnofabrics: An Example in the Uppermost Cretaceous Chalk of Denmark. *Ichnofabric and Ichnofacies*, **SEPM**, 232-249.

Exner, F.M. (1931) Zur Dynamik der Bewegungsformen auf der Erdoberfläche. *Ergebnisse der Kosmischen Physik, Akademische Verlagsgesellschaft, Leipzig, FRG*, **1**, 373-445.

Faujas St-Fond (1799) *Histoire naturelle de la montagne St-Pierre de Maestricht*. Imprimeur-Libraire H. Jansen, An 7é de la Répub. Francaise, 263 pp.[*]

Felder, P.J. (1971) Een vuursteen-rolsteen uit de Maastrichtse Kalken. *Grondboor en Hamer. Tijdschr. Nederl. Geol. Vereninging*, **4**, 88-90.

Felder, P.J. (1975) Zusamenhänge zwischen Feuerstein und dem Sediment in den Limburger Kalken aus dem Campan-Maastricht. *Tweede Intern. Sympos. over Vuursteen. Nederlandse Geol. vereninging Staringia* **3**, 21-22.

Felder, P.J. (1981) Mesofossielen in de kalkafzettingen uit het Krijt van Limburg. *Publ. Natuurhist. Genootschap in Limburg*, **31 (1-2)**, 1-35.

Felder, P.J. (1982) Mesofossils in the Cretaceous chalk of Heugem 1 and Kastanjelaan 2. *Publ. Natuurhist. Genootschap in Limburg*, **32 (1-4)**, 44-45.

Felder, P.J. (1986) Rhythms, flint and mesofossils in the Cretaceous (Maastrichtian) of Limburg, The Netherlands. In: *The Scientific Study of Flint and Chert*. (Ed. by Sieveking, G. & Hart, M. B.). pp. 83-87. Cambridge University press. 290 pp.

Felder, P.J., Bless, M.J.M., Demyttenaere, R., Dusar, M., Meessen, J.P.M.Th. & Robasynski, F. (1985) Upper Cretaceous to Early Tertiary Deposits (Santonian-Paleocene) in Northeastern Belgium and South Limburg (The Netherlands) with reference to the Campanian-Maastrichtian. *Belgische Geologische Dienst Professional Paper 1985/1*, **214**, 1-151.

Felder, P.J. (1988) Lithologic and bioclastic aspects of the Maastrichtian type area between Maastricht (the Netherlands) and Hallembaye (Belgium). In: *The Chalk District of the Euregio-Rhine*. (Ed. by Streel, M. & Bless, M.J.M.). pp. 41-55. Nat. hist. Mus. Maastr. & Lab. Pal. Univ. d'Etat Liège. 117 pp.

Felder, P.J. & Boonen, L.G.M. (1988) Gamma-ray Measurements of Upper Cretaceous to Pleistocene deposits in South Limburg (SE Netherlands) and northern Liége (NE Belgium). In: *The Chalk District of the Euregio-Rhine*. (Ed. by Streel, M. & Bless, M.J.M.). pp. 17-24. Nat. hist. Mus. Maastr. & Lab. Pal. Univ. d'Etat Liege. 117 pp.

Felder, W.M. (1971) Een bijzondere vuursteenknol. *Grondboor- Hamer*, **2**, 30-38.

Felder, W.M. (1975a) Lithostratigraphische Gliederung der Oberen Kreide. *Publ. Natuurhist. Genootschap in Limburg*, **24 (3-4)**, 1-43.

Felder, W.M. (1975b) Lithostratigraphie van het Boven-Krijt en het Dano-Montien in ZuidLimburg en het aangrenzende gebied. In: *Toelichting bij de geologische over- zichtskaarten van Nederland*, 63-72.

Fischer, A.G. (1991) Orbital Cyclicity in Mesozoic Strata. In: *Cycles and events in stratigraphy* (Ed. by Einsele, Ricken, Seilacher) pp. 48-62. Springer-Verlag Berlin- Heidelberg-New York, 955 pp.

Fitton, W.H, et al. (1829) In: *D'Archiac, Histoire des progrès de la géologie, Tome 4, I-re partie.*[*]

Fitton, W.H. (1834) Observation on part of the Low Countries and the north of France, principally near Maestricht and Aix la Chapelle. *Proc. Geol. Soc. London*, **14**, 161-164.[*]

Floris, S. (1979) Maastrichtian and Danian Corals from Denmark. In: *Cretaceous-Tertiary boundary events, symposium, I The Maastrichtian and Danian of Denmark* (ed. by Birkelund, T. & Bromley, R.G.). pp 92-95. University of Copenhagen, 210 pp.

Fournier, R.O. & Marshall, W.L. (1983) Calculation of amorphous silica solubilities at $25°$ C and $300°$ C and apparent cation hydration numbers in aqueous salt solution using the concept of effective density of water. *Geochim. Cosmochim. Acta.*, **47**, 587-596.

Francken, C., (1947) Bijdrage tot de kennis van het Boven-Senoon in Zuid-Limburg. *Med. Geol. Stichting*, **C-VI-5**, 1-148.

Froelich, P.N. Klinkhammer, G.P. Bender, M. Luedtke, N. Heath, G.R. Cullen, D. & Dauphin, P. (1979) Early oxidation of organic matter in pelagic sediments off the eastern equatorial Atlantic: suboxic diagenesis. *Geochim. Cosmochim. Acta.*, **43**, 1075-1090.

Fürsich, F.T (1979) Genesis, environments, and ecology of Jurassic hardgrounds. *Neues Jahrb. Geol. Paläont. Abh.*, **158**, 1-63.

Gale, A.S. (1989) A Milankovitch scale for Cenomanian time. *Terra Nova*, **1**, 420-425.

Garrison, R.E. (1974) Radiolarian cherts, pelagic limestones and igneous rocks in eugeosyn- clinal assemblages. *Spec. Publs Int. Ass. Sediment.*, **1**, 367-399.

Gautier D.L. & Claypool G.E. (1984) Interpretation of methanic diagenesis in ancient sediments by analogy with processes in modern diagenetic environments. In: Clastic diagenesis. (Ed. by McDonald D.A., Surdam R.C.) *Am. Assoc. Petrol. Geol. Mem.*,

37, 111-123.

Grossouvre, A. De (1908) Description des Ammonitides du Crétacé supérieur du Limbourg Belge et Hollandais et du Hainaut. *Mém. Musée Roy. Hist. Nat. Belg.*, **4**, 1-38.

Godwin Austen (1858) The quarterly journal of the geological society, vol. XIV, part 3.*

Guilcher, A. (1969) Pleistocene and Holocene sea level changes. *Earth-Sci. Rev.*, **5**, 69-97.

Gullentops, F. (1987) The Maastrichtian Sealevel Rise. *Annales de la Société géologique de Belgique*, **109**, 363-365.

Håkansson, E., Bromley, R.G. & Perch-Nielsen, K. (1974) Maastrichtian chalk of north-west Europe - a pelagic shelf sediment. In: Pelagic sediments: on land and under the sea. (Ed. by Hsü, K.J. & Jenkyns, H.C.). *Spec. Publs Int. Ass. Sediment.*, **1**, 211-233.

Hancock, J.M. (1963) The hardness of the Irish Chalk. *Irish Naturalist J.*, **14**, 157-164.

Hancock, J.M. & Kauffman, E.G. (1979) The great transgressions of the Late Cretaceous. *Journal of the Geological Society London*, **136**, 175-186.

Hansen, H.J., Gwozdz, R., Hansen, J.W., Bromley, R.G. & Rasmussen, K.L. (1986) The diachronous C/T plankton extinction in the Danish Basin. In: *Lecture Notes in Earth Sciences, Vol. 8. Global Bio-Events.* (Ed. by Walliser, O.). Springer-Verlag Berlin Heidelberg, 381-384.

Haq, B.U. Hardenbol, J. & Vail, P.R. (1987) Chronology of fluctuating sea levels since the Triassic. *Sience*, **235**. 1156-1167.

Hardenbol, J., Vail, P.R. & Ferrer, J. (1981) Interpreting paleoenvironments, subsidence history and sealevel changes of passive margins from seismic and biostratigraphy. *Oceanologica Acta. Proceed. 26 th Internat. Geol. Congr. Geology of continental margins symposium, Paris 1980*, 33-44.

Harder, H. (1978) Synthesis of iron layer silicate minerals under natuaral conditions. *Clays and Clay Minerals*, **26-1**, 65-72.

Harder, H. (1980) Synthesis of glauconites at surface temperatures. *Clays and Clay Minerals*, **28-3**, 217-222.

Harland, W.B. et al (1982) *A geologic time scale*. Cambridge Univ. Press. 131 pp.

Harland, W.B., Armstrong, R.L., Cox, A.V., Craig, L.E., Smith, A.G., Smith, D.G. (1989) A geologic time scale. Cambridge Univ. Press. 263 pp.

Hart, M.B. Bailey, H.W. Swiecicki, A. & Lakey, B.R. (1986) Upper Cretaceous flint meal faunas from southern England. In: *The Scientific Study of Flint and Chert* (Ed. by Sieveking, G. & Hart, M.B.). pp. 89-97. Cambridge University press. 290 pp.

Hart, M.B. (1987) Orbitally induced cycles in the chalk facies of the United Kingdom. *Cret. Res.*, **8**, 335-348.

Hays, J.D. Imbrie, J. & Shackleton, N.J. (1976) Variations in the Earth's orbit: pacemaker of the ice ages. *Science*, **194**, 1121-1132.

Haug, E. (1900) Les Géosynclinaux et les aires continentales. Contribution à l'étude des transgressions et régressions marines. *Bulletin de la Societé géologique de France*, **3 (28)** , 617-711.

Herrington, P.M. Pederstad, K. & Dickson, J.A.D. (1991) Sedimentology and Diagenesis of Resedimented and Rhythmically Bedded Chalks from the Elgfisk Field, North Sea

Central Graben. *Am. Ass. Petr. Geol. Bull.*, **75**, 1161-1674.

Heybroek, P., Haanstra, U. & Erdman, D.A. (1967) Observations on the geology of the North Sea area. *Proc. 7th World Petroleum Congress*, **2**, 905-916.

Hofker, J. (1955) Ontdekking van een nog niet bekende geologische formatie in Zuid-Limburg. *Natuurhist. Maandb.*, **44**.

Hofker, J. (1956) Het Onder-Paleoceen van Zuid-Limburg. *Natuurhist. Maandb.*, **45** (11-12), 132-133.

Hofker, J. (1957) Een nieuwe laag in het bovenste Krijt van Zuid-Limburg. *Natuurhist. Maandb.*, **46** (9-10), 121-123.

Hofker, J. (1966) Maestrichtian, Danian and Paleocene Foraminifera. The Foraminifera of the type-Maestrichtian in South Limburg, Netherlands, together with the Foraminifera of the underlying Gulpen chalk and the overlying calcareous sediments, the Foraminifera of the Danske kalk and the overlying greensands and clays as found in Denmark. *Paleontograph., Suppl.*, **10**, 375 pp.

Hudson, J.D. (1967) Speculations on the depth relations of calcium carbonate solutions in Recent and ancient seas. *Marine Geol.*, **5**, 473-480.

Hudson, J.D. (1977) Stable isotopes and limestone lithification. *Geol. Soc. London Q. J.*, **133**, 637-660.

Hurd, D.C. (1973) Interactions of biogenic opal, sediment and seawater in the central equatorial Pacific. *Geochim. Cosmochim. Acta*, **37**, 2257-2282.

Hyman, L.H. (1951) *The invertebrates*. McGraw-Hill, New York. 550 pp.

Iler, R.K. (1973) Colloidal Silica. 6. In: *Surface and Colloid Science*. (Ed. by Matijevic, E.) Wiley, New York. 1-100.

Iler, R.K. (1979) *Chemistry of Silica*. Wiley-Interscience, New York, 866 pp.

Imbrie, J. & Imbrie, K.P. (1979) *Ice ages: solving the mystery*. 2nd edn. Harvard Univ. Press, Cambridge.

Jäger, M. (1988) Serpulids around the Gulpen/Maastricht Formation boundary (Upper Maastrichtian) in South Limburg (The Netherlands) and adjacent belgian areas. In: *The Chalk District of the Euregio-Rhine*. (Ed. by Streel, M. & Bless, M.J.M.). pp. 69-76. Nat. hist. Mus. Maastr. & Lab. Pal. Univ. d'Etat Liège. 117 pp.

Jagt, J.W.M. (1988) Some stratigraphical and faunal aspects of the Upper Cretaceous of Southern Limburg (The Netherlands) and contigeous areas. In: *The Chalk District of the Euregio-Rhine*. (Ed. by Streel, M. & Bless, M.J.M.). pp. 25-39. Nat. hist. Mus. Maastr. & Lab. Pal. Univ. d'Etat Liège. 117 pp.

Jeletzky, J.A. (1951) Die Stratigraphie und Belemnitenfauna des Obercampan und Maastricht Westfalens, Nordwestdeutschlands und Dänemarks sowie allgemeine Gliederungsprobleme der jüngeren borealen Oberkreide Eurasiens. *Beih. geol. Jb.*, **1**, 142 pp.

Jordan, R. (1981) Sind submarine Gas- und Schlammvulcane in der Schreibkreide-Facies Nordwesteuropas Anlass für die Genese der Paramoudras? *N. Jb. Geol. Paléont. Mh.*, **7**, 419-426.

JØrgensen, N.O. (1979) Distribution of Fe, Ti, Mn, and Ba in Maastrichtian and Danian carbonate rocks in the Danish Basin and the North Sea Central Graben. In: *Cretaceous-Tertiary boundary events. Symposium. I, The Maastrichtian and Danian of Denmark*

(Ed. by Birkelund, T. & Bromley, R.G.) pp. 16-32. University of Copenhagen. 210 pp.

Kamatani, A. & Riley, J.P. (1979) Rates of dissolution of diatom silica walls in seawater. *Marine Biol.*, **55**, 29-35.

Kastner, M. Keene, J.B. & Gieskes J.M. (1977) Diagenesis of siliceous oozes. I. Chemical controls on the rate of opal-A to opal-CT transformation. An experimental study. *Geochim. Cosmochim. Acta*, **41**, 1041-1059.

Kaunhowen, F. (1897) Die Gastropoda der Maastrichter Kreide. *Pal. Abh. N. F.*, **4 (1)**, 3-132.

Kennedy, J.F. (1964) The formation of sediment ripples in closed rectangular conduits and in the desert. *Journal of Geophysical Research*, **69-8**, 1517-1524.

Kennedy, W.J. (1970) Trace fossils in the Chalk environment. In: *Trace Fossils* (Ed. by Crimes, T.P. & Harper, J.G.). *Geol. J. Spec. Iss.*, **3**, 263-282.

Kennedy, W.J. & Juignet, P. (1974) Carbonate banks and slump beds in the Upper Cretaceous (Upper Turonian-Santonian) of Haute Normandy, France. *Sedimentology*, **21**, 1-42.

Kennedy, W.J. & Garrison, R.E. (1975) Morphology and genesis of nodular chalks and hardgrounds in the Upper Cretaceous of southern England. *Sedimentology*, **22**, 311-386.

Knauth. L.P. (1979) A model for the origin of chert in limestone. *Geology*, **7**, 274-277.

Krauskopf, K.B. (1959) The geochemistry of silica in sedimentary environments. In: Silica in sediments. *Soc. Econ. Paleontol. Mineral. Spec. Publ.*, **7**, 4-19.

Kruytzer, E.M. & Meijer, M. (1958) On the occurrence of Crania brattenburgica (v. Schlotheim 1820) in the region of Maastricht (Netherlands) (Brachiopoda, inarticulata). *Natuurhist. Maandbl.*, **47**, 135-141.

Kruytzer, E.M. (1964) De Mosasauriers van ons Krijt. (Les Mosasauriens du Crétacé supérieur du Limbourg méridional, Pays-Bas). *Natuurhist. Maandbl.*, **53**, 150-157.

Kruytzer, E.M. (1969) Le genre Crania du Crétacé supérieur et du Post-Maastrichtien de la Province de Limbourg néerlandais (Brachiopoda, Inarticulata). *Publ. Natuurhist. Gen. Limburg*, **19 (3)**, 5-42.

Lambert, J. (1911) Description des Echinides de la Belgique, II, Echinides de l'Étage Santonien. *Mus. r. hist. nat. Belgique*, **16**, 1-81.

Latimer, W.M. (1952) *Oxidation potentials.* Prentice Hall, Englewood Clifs, N. J. 392 pp.

Leary, P.N. Cottle, R.A. & Ditchfield, P. (1989) Milankovitch control of foraminiferal assemblages from the Cenomanian of Southern England. *Terra Nova*, **1**, 416-419.

Legrand, R. (1961) L'épirogénése, sources tectoniques, d'après des exemples choisis en Belgique. *Mém. Inst. Géol. Univ. Louvain*, **22**, 3-66.

Legrand, R. (1968) Le Massif du Brabant. *Mem. Serv. Geol. Belg.*, **9**, 1-148.

Le Luc, (1799) Lettres physiques et morales.*

Leriche, M. (1927) Les Poissons du Crétacé marin de la Belgique et du Limbourg hollandais. Note prél. Les résultats stratigraphiques de leur étude. *Bull. Soc. géol. Belg.*, **36**, 199-299.

Lyell, C. (1854) *Manual of Elementary Geology.* John Murray London, 398 pp.*

Marsaglia, K.M. & Klein, G.D. (1983) The Paleogeography of Paleozoic and Mesozoic storm depositional systems. *The Journal of Geology*, **91**, 117-142.

Marshall, J.D. & Ashton, M. (1980) Isotopic and trace element evidence for submarine lithification of hardgrounds in the Jurassic of eastern England. *Sedimentology*, **27**, 271-291.

Mathieu, (1813)*

McElhinny, M. (1973) *Palaeomagnetism and plate tectonics*. Cambridge Univ. Press London.

McLean, D.M. (1985) Deccan Traps Mantle Degassing in the Terminal Cretaceous Marine Extinctions. *Cretaceous Res.*, **6**, 235-259.

Meessen, J.P.M.Th. (1977) Foraminiferen onderzoek van enige monsters van het Onder Tertiar en Boven Krijt van drie diepboringen uit Noord-oost België. *Prof. Paper. Ministerie van Economische Zaken, aardkundige Dienst van België*, **1**, 1-5.

Meijer, M. (1959) Sur la limite supérieure de l'étage Maastrichtien dans la région type. *Bull. Acad. r. Belge*, **45**, 316-338.

Meijer, M. (1965) The stratigraphical distribution of Echinoids in the Chalk and Tuffaceous Chalk in the neighbourhood of Maastricht (Netherlands). *Meded. geol. Sticht. N. S.*, **17**, 21-25.

Muller, J.E. (1945) De Post-Carbonische tektoniek van het Zuid-Limburgse mijngebied. *Meded. Geol. Stichting*, **C-I-2**, 32.

Nestler, H. (1965) Die Rekonstruktion des Lebensraumes der Rügener Schreibkreide-Fauna (Unter-Maastricht) mit Hilfe der Paläoökologie und Paläobiologie. *Geologie*, **14**, 147 pp.

Neugebauer, J. (1974) Some aspects of cementation in chalk. In: Pelagic Sediments: on Land and under the Sea (ed. by Hsü, K.J. & Jenkyns, H.C.). *Spec. Publs int. Ass. Sediment.*, **1**, 177-210.

Nygaard, E. (1983) Bathichnus and its significance in the trace fossil association of Upper Cretaceous chalk, Mors, Denmark. *Danm. geol. Unders. Arborg*, 107-137.

Odin, G.S. & Matter, A. (1981) De glauconiarum origine. *Sedimentology*, **28**, 611-641.

Okamoto, G. Okura, T. & Goto, K. (1957) Properties of silica in water. *Geochim. Cosmochim. Acta*, **12**, 123-132.

Omalius d'Halloy, J.J.D' (1822) *Essai d'une carte géologique des Pays-Bas, de la France et de quelques contrés voisins*. Mém. géolog. Namur,.*

Omalius d'Halloy, J.J.D' (1848) *Coup d'oeil géologique de la Belgique*. Bruxelles.*

Paasche, E. (1968) Biology and physiology of coccolithophorids. *Annu. Rev. Microbiol.*, **22**, 77-86.

Parks, G.A. (1967) Aquaous surface chemistry of oxides and complex oxide minerals. *Am. Chem. Soc. Adv. Chem. Ser.*, **67**, 121-160.

Patijn, R.J.H. (1961) Bewegingen langs de Heerlerheide- en Geleen-Storing in het Mauritsveld. *Meded. Geol. Sticht.*, **3**, 35-38.

Patijn, R.J.H. & Kimpe, W.F.M. (1961) De kaart van het Carboon-oppervlak, de profielen en de kaart van het dekterrein van het Zuid-Limburgs mijngebied en Staatsmijn Beatrix en omgeving. *Med. Geol. Stichting*, **C-I-1(4)**, 1-12.

Pomel (1848)*

Priem, H.N.A., Boelrijk, N.A., Hebeda, E.H., Romein, E.A. & Verschure, R.H. (1975)

Biotopic dating of Glauconites from the upper Cretaceous in the Netherlands and Belgian Limburg. *Geol. en Mijnb.*, **54**, 205-207.

Purser, B.H. (1969) Syn-sedimentary marine lithification of Middle Jurassic limestones in the Paris Basin. *Sedimentology*, **12**, 205-230.

Pyzic, A.J. & Sommer, S.E. (1981) Sedimentary iron mono-sulphides: Kinetics and mechanisms of formation. *Geochim. Cosmochim. Acta*, **45**, 687-689.

Raiswell, R. (1987) Non-steady state microbiological diagenesis and the origin of concretions and nodular limestones. In: *Diagenesis in sedimentary sequences* (Ed. by Marshall, M.D.) Geol. Soc. London, Blackwell, Oxford, pp. 41-54.

Rampino, M.R. (1982) A non-catastrophist explanation for the Iridium anomaly at the Cretaceous/Tertiary boundary. *Geol. Soc. Am. Special Paper*, **190**, 455-460.

Rasmussen, H.W. (1971) Echinoid and crustacean burrows and their diagenetic significance in the Maastrichtian-Danian of Stevns-Klint, Denmark. *Lethaia*, **4**, 191-216.

Rasmussen, L.B. (1978) Geological aspects of the Danish North Sea Sector. *Danmarks geol. Unders.*, **3**, 4-85.

Raudkivi, A.J. & Witte, H.H. (1990) Development of Bed Features. *Journal of hydraulic engineering*, **116**, 1063-1079.

Ravn, J.P.J. (1925) Sur le placement géologique du Danien. *Danmarks geol. Unders.*, **2**, 1-48.

Reeburgh, W.S. (1980) Anaerobic methane oxidation: rate depth distributions in Skan Bay sediments. *Earth Plat. Sci. Lett.*, **47**, 345-352.

Redfield, A.C. Ketchum, B.H. & Richards F.A. (1963) The influence of organisms on the composition of seawater. In: *The Sea. 2.* (Ed. by M.H. Hill). Wiley Intersience, New York, 26-77.

Reid, R.E.H. (1968) Bathymetric distribution of Calcarae and Hexactinellida in the present and the past. *Geol. Mag.*, **105**, 546-559.

Renier, A. (1902) Le poudingue de Malmédy, Essay géologique. *Annales de la Société géologique Belgique*, **29 M**, 145-223.

Richards, K.J. (1980) The formation of ripples and dunes on an erodible bed. *Journal of Fluid Mechanics*, **99-3**, 597-618.

Robaszynski, F. (1981) Moderation of Cretaceous Transgressions by Block Tectonics. An Example from the North-West of the Paris Basin. *Cretaceous Research*, **2**, 197-213.

Robaszynski, F. (1988) Upper Cretaceous planktonic foraminifera from northern Belgium and the southeastern Netherlands. In: *The Chalk District of the Euregio-Rhine.* (Ed. by Streel, M. & Bless, M.J.M.). pp. 77-83. Nat. hist. Mus. Maastr. & Lab. Pal. Univ. d'Etat Liege. 117 pp.

Roemer, F.A. (1840) Die Versteinerungen des Norddeutchen Kreidegebirges.˙

Romein, B.J. (1962) On the typelocality of the Maastrichtian (Dumont, 1849), the upper boundary of that stage and on the transgression of the Maastrichtian s. l. in South Limburg. *Guide to the excursion D of the jubilee convention*, **14**.

Romein, B.J. (1963) Present knowledge of the stratigraphy of the Upper Cretaceous (Campanian-Maastrichtian) and Lower Tertiary (Dano-Montian) calcareous sediments in southern Limburg. *Med. Geol. Stichting, N. S.*, **15**, 77-84.

Rossa, H.G. (1987) Upper Cretaceous and Tertiary inversion tectonics in the western part of the Rheinish-Westphalian coal district (FRG) and in the Campine area (N. Belgium). *Ann. Soc. géol. Belg.*, **109**, 367-410.

Roy, A.B. & Trudinger, P.A. (1970) *The Biochemistry of inorganic Compounds of Sulfur*. Cambridge University Press.

Savrda, C.E., Bottjer, D.J. & Seilacher, A. (1991) Redox-Related Benthic Events. In: *Cycles and events in stratigraphy* (Ed. by Einsele, Ricken, Seilacher) pp. 542-542. Springer-Verlag Berlin-Heidelberg-New York, 955 pp.

Schins, W.J.H. & Buurman, P. (1979) Silicification phenomena in fossil Belemnite guards. *Derde Intern. Sympos. over Vuursteen. Nederlandse Geol. vereninging Staringia*, **6**, 22-25.

Schlanger, S.O. & Douglas, R.G. (1974) The pelagic ooze-chalk-limestone transition and its implications for marine stratigraphy. In: Pelagic Sediments: on Land and under the Sea. (Ed. by Hsü, K.J. & Jenkyns, H.C.). *Spec. Publs int. Ass. Sediment.*, **1**, 117-148.

Schlegel, H.G. (1992) *General Microbiology*. Cambridge University Press. pp 587.

Schmid, F. (1959) Biostratigraphie du Campanien-Maastrichtien du NE de la Belgique sur la base des Bélemnites. *Ann. Soc. géol. belg.*, **82 B**, 235-256.

Scholle, P.A. (1971) Sedimentology of fine-grained deep-water carbonate turbidites. Monte Antola Flysch (Upper- Cretaceous) Northern Apennines, Italy. *Geol. Soc. Amer. bull.*, **82**, 629-658.

Scholle, P.A. (1974) Diagenesis of Upper Cretaceous chalks from England, Northern Ireland, and the North Sea. In: Pelagic Sediments: on Land and under the Sea. (Ed. by Hsü & Jenkyns). *Spec. Publs int. Ass. Sediment.*, **1**, 177-210.

Schwarzacher, W. (1989) Milankovitch cycles and the measurement of time. *Terra Nova*, **1**, 405-408.

Séronie-Vivien, M. (1972) Contribution a l'étude du Sénonien en Aquitaine Septentrionale. *Éditions du centre national de la recherche scientifique*. 195 pp.

Siever, R. (1962) Silica solubility, O-200° C, and the diagenesis of siliceous sediments. *J. Geol.*, **70**, 127- 150.

Siever, R. & Woodford, N. (1973) Sorption of silica by clay minerals. *Geochim. Cosmochim. acta*, **37**, 1851-1880.

Simien, T.R. (1987) From coastal terrigenous deposits to pelagic rich oozes: the Holocene transgressive complex of Southern Belize Lagoon (C. A. ) A tool for the past? *Comparative sedimentol. Lab. Univ. Miami, Rosenstiel School Marine and Atmospheric Sciences, Miami*, 32-35.

Smiser, J.S. (1935) A monograph of the Belgian Cretaceous Crinoids. *Mém. Mus. r. hist. nat. Belgique, 1. s.*, **68**, 1-98.

Smit, J. & Hertogen, J. (1980) An extraterrestrial event at the Cretaceous-Tertiary boundary. *Nature*, **285**, 198-200.

Smit, J. & Ten Kate, W.G.H.Z. (1982) Trace-element patterns at the Cretaceous-Paleogene boundary -- consequences of a large impact. *Cretaceous Research*, **3**, 307-332.

Soudry, D., Moshkovitz, S. & Ehrlich, A. (1981) Occurrence of siliceous microfossils (diatoms, siliciflagellates and sponge spicules) in the Campanian Mishash Formation, Southern Israel. *Eclog. Geol. Helv.*, **74**, 97-107.

Streel, M. & Bless, J.M. (1988) The Chalk district of the Euregio Meuse-Rhine. Selected papers on Upper Cretaceous deposits. Natuurhistorisch Museum Maastricht and the Laboratoires de Paléonthologie de l'Université d'Etat de Liège, 1-117.

Steinich, G. (1965) Die artikulaten Brachiopoden der Rügener Schreibkreide (Unter-Maastricht). *Palaeontographica*, **A2** (1), 220 pp.

Stenestad, E. (1972) Troek af det danske bassins udvikling i Øvre Kridt. *Danske geol. Foren. Arsskrift for 1971*, 63-69.

Stresemann, E. (1970) *Excursionsfauna von Deutschland. Wirbellose 1.* Volk und Wissen, Berlin. I-XXXIV. 494 pp.

Suess, E. (1900) *Das Antlitz der Erde (Band II).* Prague.

Suess, E. (1979) Mineral phases formed in anoxic sediments by microbial decomposition of organic matter. *Geochim. Cosmochim. Acta*, **43**, 339-352.

Surlyk, F. (1970) Die Stratigrapie des Maastricht von Dänemark und Norddeutshlands auf grund von Brachiopoden. *Newsl. Stratigr.*, **1**, 7-16.

Surlyk, F. (1972) Morphological adaptations and population structures of the Danish Chalk Brachiopodes (Maastrichtian, Upper Cretaceous). *Biol. Skr.*, **19**, 57 pp.

Surlyk, F. (1979) Guide to Stevns Klint. *Cretaceous-Tertiary Boundary Events. Symposium I. The Maastrichtian and Danian of Denmark* (Ed. by Birkelund, T. & Bromley, R.G.). pp. 164-170. University of Copenhagen, 210 pp.

Swinchatt, J.P. (1965) Algal boring: a possible depth indicator in carbonate rocks and sediments. *Geol. Soc. Am. Bull.*, **80**, 1391-1396.

Takahashi, J.I. (1939) Synopsis of glauconitization. In: *Recent marine sediments, a symposium* (Ed. by Trask, P.D.), pp. 503-512, Tulsa.

Ten Kate, G.H.Z. & Sprenger, A. (1993) Orbital cyclicities above and below the Cretaceous / Paleogene boundary at Zumaya (N Spain), Agost and Relleu (SE Spain). *Sedimentary Geology*, **87**, 69-107.

Thomsen, E. (1976) Depositional environment and development of Danian bryozoan biomicrite mounds (Karlby Klint, Denmark). *Sedimentology*, **23**, 485-509.

Thorez, J. & Monjoie, A. (1972) Lithologie et assemblages argileux de la smectite de Herve et des craies Campaniennes et Maestrichtiennes dans le Nord-Est de la Belgique. *Ann. Soc. geol. Belg.*, **96**, 651-670.

Triger, (1857)*

Ubachs, J.C. (1858) Neue Bryozoen-Arten aus der Tuff-Kreide von Maestricht. *Paleontographica*, **5**, 127-131.

Ubachs, C. (1879) *Description géologique et paléontologique du sol du Limbourg.* Ruremonde, 275 pp.

Uhlenbroek, G.D. (1912) Het Krijt van Zuid-Limburg. Toelichting bij eene geologische kaart van het Krijt-gebied van Zuid- Limburg. *Jaarverslag Rijksopsporingsd. van delfstoffen over 1911*, 48-57.

Umbgrove, J.H.F. (1925) De Anthozoa uit het Maastrichtsch tufkrijt. *Leidsche geol. Meded.*, **I**, 83-126.

Umbgrove, J.H.F. (1926) Bijdrage tot de kennis der Stratigraphie, Tektoniek en Petrographie van het Senoon in Zuid-Limburg. *Leidsche geol. Mededeel.*, **2**, 255-332.

Umbgrove, J.H.F. (1927) Over Lithothamnia in het Maastrichtse tufkrijt. *Leidsche geol. Meded.*, **II**, 89-97.

Vail, P.R., Mitchum, R.M. & Thomson, S. III (1977) Seismic stratigraphy and global changes of sealevel, part 3: Relative changes of sealevel from coastal onlap. In: *Seismic stratigraphy - applications to hydrocarbon exploration* (Ed. Payton, P.C.). *Am. Assoc. Petrol. Geolog. Memoir*, **26**, 63-82.

Van De Geijn, W.A.E. (1937) Les élasmobranches du Crétacé marin du Limbourg hollandais. *Nat. hist. maandbl.*, **XXVI**, 16-21, 28- 32, 56-69.

Van De Geijn, W.A.E. (1940) Les Rudistes du tuffeau de Maestricht (Sénonien supérieur). *Nat. Hist. Maandbl.*, **XXIX**, 51-57.

Van Der Weijden, C.H. Middelburg, J.J. & Gaans, van. P.L.M. (1989) Comment on Zijlstra, J.J.P. (1987) *Geol. Mijnb.*, **68**, 263-270.

Van Gemerden, H. (1993) Microbial mats: A joint venture. *Marine Geology*, **113**, 3-25.

Vangerow, E.F. & Schloemer, W. (1967) Vergleich des "Vetsschauwer Kalkes" der Aachener Kreide mit dem Kreide-Profil von Süd-Limburg anhand von Coccolithen. *Geol. Mijnb.*, **46**, 453-458.

Vanguestaine, M. (1966) Étude palynologique quantitative des deux carrieres du Crétacé superieur de la valée de la Meuse. *Bull. Cl. Sci. Acad. R. Belg.*, **(5) 52**, 1534-1548.

Van Harten, D. (1972) Heavy minerals in the Maastrichtian type-region. *G.U.A. Papers gel., Netherl.*, **1**, 1-85.

Van Hinte, J.E. (1976) A Cretaceous Time Scale. *Amer. Ass. Petrol. Geol. Bull.*, **60/4**, 498-516.

Van Veen, J.E. (1928) Vorläufige Mitteilung über die Cytherella Arten der Maastrichter Tuffkreide. *Natuurhist. Maandbl.*, **17**, 123-125.

Van Veen, J.E. (1932) Die Cytherellidae der Maastrichter Tuffkreide und des Kunrader Korallenkalkes von Süd-Limburg. *Verh. k. nederl. geol. mijnbouwkd. Genoot. Geol. Ser.*, **IX**, 317- 364.

Van Veen, J.E. (1934) Die Cypridae und Bairiidae der Maastrichter Tuffkreide und des Korrallenkalkes von Süd-Limburg. *Natuurhist. Maandbl.*, **23**, 88-132.

Van Veen, J.E. (1935-1936) Die Cytheridae der Maastrichter Tuffkreide und des Kunrader Korrallenkalkes von Süd-Limburg. *Natuurhist. Maandbl.*, **24**, 26-112, **25**, 21-170. Maastricht.

Van Veen, J.E. (1936) Nachtrag zu der bis jetzt erschienenen Revision der Ostracoden der Maastrichter Tuffkreide und des Korrallenkalkes von Süd-Limburg. *Natuurhist. Maandblad*, **25**, 170-187.

Villain, J.M. (1975) "Calcisphaerulidae" (Incertae sedis) du Cretace superieur du Limbourg. (Pays-Bas) et d'autres regions. *Paleontographica (A)*, **149**, 193-242.

Villain, J.M. (1977) Le Maastrichtien dans sa région type (Limbourg, Pays-Bas) étude stratigraphique et micro-paléonthologique. *Paleonthographica (A)*, **157**, 1-87.

Voigt, E. (1929) Die Lithogenese der Flach- und Tiefwasser-sedimente des jüngern Oberkreidemeeres. *Jb. Halle. Verb.*, **8**, 1-165.

Voigt, E. (1959) Die ökologische bedeutung der Hartgründe ("Hardgrounds") in der oberen Kreide. *Paleont. Z.*, **33** (3), 129-147.

Voigt, E. (1968) Über Hiatus-Konkretionen, dargestellt an Beispielen aus dem Lias. *Geol. Rdsch.*, **58**, 281-296.

Voigt, E. (1974) Über die Bedeutung der Hartgründe (Hardgrounds) für die

Evertebratenfauna der Maastrichter Tuffkreide. *Natuurhist. Maandbl.*, **63 (2)**, 32-39.

Voigt, E. (1979a) Über die Zeit der Bildung der Feuersteine in der Oberen Kreide. *Derde Intern. Sympos. over Vuursteen. Nederlandse Geol. vereninging. Staringia*, **6**, 11-16.

Voigt, E. (1979b) Kritische Bemerkungen zur Diskussion über die Kreide/Tertiärgrenze. In: *Cretaceous-Tertiary boundary events. Symposium. I, The Maastrichtian and Danian of Denmark* (Ed. by Birkelund, T. & Bromley), pp. 16-32. R.G. University of Copenhagen. 210 pp.

Voigt, E. & Domke, W. (1955) Thallassiocharis bosqueti Debey & Miquel ein structurel erhaltenes Seegras aus der holländischen Kreide. *Mitt. geol. St. Inst. Hamb.*, **24**, 87-102.

Von Köster, H.M. & Kohler, E.E. (1973) Sediment petrographische und mineralogische Untersuchungen an glaukonitführenden Kreidegesteinen Bayerns. *Geologischen Rundschau*, **62**, 521-535.

Wegener, A. (1915) *Die Entstehung der Kontinente und Ozeane*. Braunschweig, Vieweg.

Wienberg Rasmussen, W. (1965) The Danian affinities of the Tuffeau de Ciply in Belgium and the "Post-Maastrichtian" in the Netherlands. *Meded. geol. Sticht. N. S.*, **17**, 33-40.

Williams, L.A. Parks, G.A. & Crerar, D.A. (1985) Silica diagenesis, I. Solubility controls. *J. Sediment Petrol.*, **55**, 301-311.

Williams, L.A. & Crerar, D.A. (1985) Silica diagenesis, II. General mechanisms. *J. Sediment Petrol.*, **55**, 312-321.

Wilson, M.A. (1990) *Studies of Inorganic Marine Hard Substrates (A Bibliography)*. Geology Dep., College of Wooster, Ohio, USA. 57 pp.

Winterer, E.L., Ewing, J.L. et al. (1973) *Init. Rep. Deep Sea Drill. Proj.* **8**, 1756 pp.

Wise, S.W. & Weaver, F.M. (1974) Chertification of oceanic sediments. In: *Pelagic Sediments: on Land and under the Sea* (Ed. by Hsü, K.J. & Jenkyns, C.J.). *Int. Ass. Sediment Spec. Publ.*, **1**. 301-326.

Woese, C.R. (1981) Archaebacteria. *Sci. Am.*, **244 6**, 98-106.

Wolfe, M.J. (1968) Lithification of a carbonate mud: Senonian Chalk in Northern Ireland. *Sediment. Geol.*, **2**, 263-290.

Wolfram, S. (1986) *Theory and applications of cellular automata*. World Scientific, Singapore. 560 pp.

Wood, C.J. & Smith, E.G. (1978) Lithostratigraphical classification of the Chalk in North Yorkshire, Humberside and Lincolnshire. *Proc. Yorks. geol. Soc.*, **42**, 263-287.

Worm, O. (1655) *Museum Wormianum*. Amsterdam, Elsevier. 389 pp.

Wright, L.D., Boon J.D. III, Green M.O., List J.H. (1986) Response of the mid shoreface of the southern mid-Atlantic bight to a "northeaster". *Geo-Mar Lett*, **6**, 153-160.

Zemmels, I. & Cook, H.E. (1973) X-ray mineralogy of sediments from the Central Pacific Ocean. In: *Initial Reports of the Deep Sea Drilling Project* (Ed. by Winterer, L. et al.), **XVII**, 517-559.

Ziegler, P.A. (1982) *Geological Atlas of Western and Central Europe*. Elsevier, New York, Amsterdam. 130 pp.

Zijlstra, J.J.P. (1987) Early diagenetic silica precipitation, in relation to redox boundaries and bacterial metabolism in late Cretaceous chalk of the Maastrichtian type locality. *Geol. Mijnb.*, **66**, 343-355.

Zijlstra, J.J.P. (1989) Reply on van der Weijden et al. *Geol. Mijnb.*, **68**, 263-270.

# Subject Index

# Lecture Notes in Earth Sciences

# Springer-Verlag
# and the Environment

We at Springer-Verlag firmly believe that an international science publisher has a special obligation to the environment, and our corporate policies consistently reflect this conviction.

We also expect our business partners – paper mills, printers, packaging manufacturers, etc. – to commit themselves to using environmentally friendly materials and production processes.

The paper in this book is made from low- or no-chlorine pulp and is acid free, in conformance with international standards for paper permanency.